COMPUTERIZED QUALITY CONTROL:
Programs for the analytical laboratory

ELLIS HORWOOD SERIES IN ANALYTICAL CHEMISTRY
Series Editors: Dr. R. A. CHALMERS and Dr. MARY MASSON, University of Aberdeen
Consultant Editor: Prof. J. N. MILLER, Loughborough University of Technology

COMPUTERIZED QUALITY CONTROL: Programs for the analytical laboratory

T. F. HARTLEY, B.Sc.(Hons.), Ph.D.
Principal Hospital Scientist
Division of Clinical Chemistry
The Institute of Medical and Veterinary Science
Adelaide, Australia

ELLIS HORWOOD LIMITED
Publishers · Chichester

Halsted Press: a division of
JOHN WILEY & SONS
New York · Chichester · Brisbane · Toronto

First published in 1987 by
ELLIS HORWOOD LIMITED
Market Cross House, Cooper Street,
Chichester, West Sussex, PO19 1EB, England
The publisher's colophon is reproduced from James Gillison's drawing of the ancient Market Cross, Chichester.

Distributors:

Australia and New Zealand:
JACARANDA WILEY LIMITED
GPO Box 859, Brisbane, Queensland 4001, Australia

Canada:
JOHN WILEY & SONS CANADA LIMITED
22 Worcester Road, Rexdale, Ontario, Canada

Europe and Africa:
JOHN WILEY & SONS LIMITED
Baffins Lane, Chichester, West Sussex, England

North and South America and the rest of the world:
Halsted Press: a division of
JOHN WILEY & SONS
605 Third Avenue, New York, NY 10158, USA

© 1987 T.F. Hartley/Ellis Horwood Limited

British Library Cataloguing in Publication Data
Hartley, T.F.
Computerized quality control: Programs for the analytical laboratory. —
(Ellis Horwood series in analytical chemistry)
1. Chemistry, Analytic — Quality control — Data processing
I. Title
543   QD75.4.E4

Library of Congress Card No. 86–25021

ISBN 0–85312–964–9 (Ellis Horwood Limited)
ISBN 0–470–20761–2 (Halsted Press)

Phototypeset in Times by Ellis Horwood Limited
Printed in Great Britain by R.J. Acford, Chichester

Contents

To Donna for her patience and interest and to
James and Emma for their patience and
curiosity while I 'word processed'.

Acknowledgements

The author would like to acknowledge the assistance of Dr David Thomas, Senior Director of the Division of Clinical Chemistry at the Institute of Medical and Veterinary Science, Adelaide, Australia, for making available the computing facilities necessary to develop the programs described in this book.

In addition my thanks are due to my colleagues whose constant enquiries on 'matters statistical' have provided me with the stimulus to finally document some of the theory, advice and programs. Ted Huber deserves special mention because it was he who focussed our attentions upon the correct approaches to calibration curve fitting, quality control and statistics. Valuable technical support was provided by Rick Tormet and Don Ide.

March 1986

T. F. Hartley

Introduction

Every analytical chemist would like to believe that the data reported from their laboratory are of a uniform and sustained quality. However, it is inevitable that from time to time doubt about some of the batches creeps in. The reasons for such doubts can originate from a number of sources. One may be that the usual method has been used to analyse a batch of specimens with a matrix that was slightly different from that for which the method was optimized. Consequently there is a real possibility that there may be a constant or systematic bias throughout the data reported for those samples. Will this possible bias affect the customer's perception of what the results mean? Another more common suspicion is that the instrumentation is either not operating or being operated exactly as intended. There are other examples, so clearly there is a need for some objective assistance in this sensitive and critical area of the perceived versus the actual analytical performance of the laboratory.

In this book I have aimed to outline automated statistical techniques that can provide this type of assistance without making unreasonable demands on the analyst's valuable time. These methods are all based on well accepted quality control procedures that can be run easily as BASIC computer programs on a personal computer or computer terminal in the laboratory. In addition I have included computerized procedures to assist the analyst with the selection of analytical performance criteria that will ensure that their customers are provided with results at the level of accuracy and precision that they require.

The approach has been from the point of view of a practising analyst who would prefer to be provided with the essential information required, with the option to follow up the full theoretical details when convenient. For this reason the bibliography and references extend beyond those items cited specifically in the text. The reader is urged to study all the titles in that section and decide which of those would be particularly relevant to their organization's activities. One feature that may be noticed while consulting the Bibliography covers a considerable time span; from 1940 to the present. It would appear that although most statistical quality control techniques were developed some years ago it has taken some time for them to be implemented on an extensive basis in analytical laboratories. This is a reflection of a common attitude to quality control which is that it is tedious and therefore only the minimum necessary will be put into everyday use. It is to be hoped that computerized quality control will shield the user from the tedium, encourage analysts to adopt more than just the minimum and eventually extend their interest to the more sophisticated techniques available.

The programs have all been written in Personal BASIC, (Digital Research). They were developed on a concurrent CPM–86 v. 3.1 computer (Labtam 3000) with a 132 column dot matrix printer. (Two programs require more than 80 columns to accommodate the printout conveniently but these could be adapted to an 80 column format if necessary.) The BASIC words actually used are listed below; it will be evident that no uncommon features of the language have been implemented

ABS	AND	CLOSE	DIM	ELSE	EXP
FOR	GOSUB	GOTO	IF	INKEY$	INPUT
INPUT#	INT	LOG	LPRINT	MID$	NEXT
OPEN	OR	PRINT	PRINT#	READ	REM
RETURN	RND	SQR	STEP	STOP	STRING$
TAB	THEN	TO	VAL	WIDTH	WRITE#
+	–	/	*	=	>=
<=	^				

The programs have been written as far as possible in an open, readable format, so there are some inefficiencies in their structure. This was unavoidable because I felt that it was more important to write straightforward code. They should be read as integral parts of the text. Their organization follows as closely as possible the accompanying description of the technique and the equations. Memory requirements for the programs and data files are modest in comparison with contemporary commercial software packages:

PROGRAM NAME	K BYTES
GAUSSGEN	1.3
LINCALIB	7.8
SPLINE	11.7
CURVEFIT	10.9
OCTABLE	2.2
OCCURVE	3.7
GMREG	5.0
QCNAME	1.3
QCFILER	1.9
VMASKA	8.3
REPORT	7.6
TOTAL	61.7

Data files require 1.5 K Bytes each.

Overall I have aimed to provide a relevant resource of fully documented computer programs which can be used, as the modules, for an objective quality-control scheme suitable for use throughout most analytical laboratories.

The programs presented in this book are available on floppy disk from the publishers: Ellis Horwood Ltd., Market Cross House, Cooper Street, Chichester, West Sussex, PO19 1EB, U.K. Several disk formats will be available, including IBM–PC and Apple II; please contact the publishers for further details.

1

Calibration graphs

1.1 INTRODUCTION

The analytical method requires that at an early stage in the procedure, the analyst must make some reference to the results of the analysis of a series of standard materials. Quality control should start here, and we therefore have included an extensive discussion on the computerized production of calibration graphs with appropriate statistical tests included to assist with the objective assessment of their quality. If the calibration data fail these tests, it is pointless to proceed with the batch of analyses. Because it is known, and is therefore the 'independent variable', the amount of analyte in the standard is plotted along the abscissa and the instrumental response, e.g. absorbance, counts per minute, electrode potential, is plotted on the ordinate. The shape of the calibration graph depends on the nature of the analytical system: it may be:

1. A straight line with a positive or negative slope.
2. A curve that is convex with respect to the abscissa.
3. A curve that is concave with respect to the abscissa.
4. A curve that is concave with respect to both the ordinate and the abscissa.
5. A curve that is convex with respect to both the ordinate and the abscissa.
6. A sigmoid curve with either a positive or negative slope.

1.2 THE LINEAR CALIBRATION GRAPH

A straight-line calibration graph is the form preferred by most analysts because of its well defined statistical properties. Least-squares linear regression analysis is now a well established technique in analytical laboratories, partly as a result of the ready availability of hand-held calculators with a suitable program in the ROM. Unfortunately this easy access to linear-regression analysis has led to its uncritical use. The program we have devised goes beyond the simple derivation of the slope, intercept and correlation coefficient and provides the user with information on the quality of the fit of the calculated linear regression equation to the original calibration data from which it was derived. The program executes the following sequence:

PARTS A and B
Enter the data into the arrays X(N) and Y(N) where N is the number of X,Y data pairs. X = the concentration in the standards and Y = the instrumental response (lines 330–332). Check the data for input errors and correct if necessary (lines 340–348).

PART C
This gives the user the option of a low resolution plot of the data on an 80 column printer (line 360–362). If this is required then a string array A$(41 rows by 61 columns) is set up and the data points are placed into the appropriate positions in this array (lines 418–434). The graph is printed out by lines 452–456.

PART D
Calculation of the intermediate results required for the determination of the regression equation; y regressed on x:

> SX = sum of the x data
> SY = sum of the y data
> XY = sum of the products of $x.y$
> XX = sum of the products of $x.x$
> YY = sum of the products $y.y$
> (lines 478–490).

PART E
Calculation of the slope and intercept with their standard deviations and the correlation coefficient.

XM   = SX/N = mean of the x data; line 498
YM   = SY/N = mean of the y data; line 500

SL   = slope of the line, y on x, line 502
      = $(XY-N.XM.YM) / (XX-N.XM\wedge 2)$

IN   = intercept of the line on the y axis; line 504
      = $YM-SL.XM$

R   = Correlation coefficient; line 506
      = $(XY-N.XM.YM/SQR[(XX-N.XM\wedge 2)(YY-N.YM\wedge 2)]$

DX   = Standard deviation of the x data; line 508
      = $SQR(N/(N-1)).SQR((XX/N)-(SX/N)\wedge 2)$

DY   = Standard deviation of the y data; line 510

SS   = Standard deviation of the estimate of the slope; line 512
      = $SQR(((DY/DX)\wedge 2).((1-R\wedge 2)/(N-2)))$

SI   = Standard deviation of the estimate of the intercept; line 514
      = $SQR(SS\wedge 2.XX/N)$

EY   = Standard error of an estimate of y from a given value of x; line 516
      = $DY.SQR(1-R\wedge 2)$

EX   = Standard error of an estimate of x from a given value of y; line 518

PART F
Printout of the results of the linear-regression analysis; each to four decimal places (lines 526–574). Checks are made at lines 548 and 568 to determine if a hardcopy printout of the results has been requested. Results displayed and/or printed include:

> Slope and standard deviation of the slope
> Intercept and standard deviation of the intercept
> Correlation coefficient
> Standard error of an estimate of y from a given value of x
> Standard error of an estimate of x from a given value of y

PART G
Comparison of the original y data with the values of y calculated from the regression equation presented in table form with the columns displaying :

> Observed y
> Calculated y for the corresponding value of x, C(N); line 592
> Difference between the observed y and calculated y, D(N); line 594
> Difference between observed y and calculated y

divided by the standard error of the estimate of y
from x, $D(N)/EY = T(N)$; line 596. Any value of $T(N)$
that is greater than 1.96 is highlighted with an
asterisk to indicate that that point is a candidate
outlier from the line of best fit; line 598

PART H
Calculation of the residual variance as a percentage of the total
variance in the y data:

$$RV = DY^2 - DY^2.R^2$$
$$RV\% = 1 - R^2; \text{ line 610.}$$

If $RV\% <= 5\%$ and $> 1\%$ then the fit is described by the program
being 'average'; line 614.
If $RV\% <= 1\%$ and 0.1% then the fit is described as 'good'; line 616.
If $RV\% <= 0.1\%$ then the fit is described as 'excellent', line 618.

PART I
This section gives the user the opportunity to recalculate the line of
best fit after excluding points identified as candidate outliers by
Section G (lines 626–656).

PART J
This section gives the user the opportunity to calculate values of x for
user entered values of y (lines 668–690).

1.2.1   Listing of LINCALIB

```
300 ' THE LINEAR CALIBRATION CURVE
302 ' PROGRAM USING LEAST SQUARES LINEAR REGRESSION ANALYSIS TO
DETERMINE THE LINE OF BEST FIT
304 ' T F HARTLEY
306 ' COMPUTERIZED QUALITY CONTROL
308 ' PUBLISHED BY ELLIS HORWOOD, ENGLAND, 1986
310 ' CLEAR SOME SCREEN
312 FOR I = 1 TO 10 : PRINT : NEXT I
314 PRINT STRING$(70, "*")
316 PRINT :PRINT "LINEAR CALIBRATION CURVE :" : PRINT "LINE OF "; :
PRINT "BEST FIT BY LEAST SQUARES LINEAR REGRESSION ANALYSIS" :
PRINT
318 PRINT STRING$(70,"*") : PRINT
320 '
322 ' PARTS A AND B ::::::::::::::::::::::::::::::::::::::::::::::
324 '
326 INPUT "HOW MANY DATA PAIRS ARE USED IN YOUR CALIBRATION CURVE "
; N : PRINT
328 DIM X(N), Y(N), C(N), D(N), T(N), A$(41), P$(1), T$(1)
330 PRINT " X = CONC IN THE STANDARD"
332 PRINT " Y = CORRESPONDING READING OBTAINED FROM THE"; :
```

```
PRINT " ANALYTICAL INSTRUMENT" : PRINT
334 FOR I = 1 TO N
336 PRINT "DATA PAIR NO. ";I; : PRINT TAB(20);
338 INPUT "X, Y = "; X,Y
340 PRINT "DATA PAIR OK ..... Y/N "
342 Q$ = INKEY$
344 IF Q$ = "Y" THEN X(I) = X : Y(I) = Y : Q$="" : GOTO 350
346 IF Q$ = "N" THEN PRINT " CORRECT " : Q$ = "" : GOTO 338
348 GOTO 342
350 NEXT I
352 '
354 ' PART C ::::::::::::::::::::::::::::::::::::::::::::::::::::
356 '
358 Q$ = ""
360 PRINT : PRINT : PRINT "DO YOU REQUIRE A HARDCOPY PRINTOUT "; :
INPUT " OF YOUR CALIBRATION GRAPH .. Y/N ";Q$
362 IF Q$= "N" THEN GOTO 474
364 PRINT : PRINT "IS THE PRINTER READY ... Y/N PROGRAM WILL "; :
INPUT "PROBABLY HALT IF NOT ! "; Q$
366 IF Q$ < "Y" THEN GOTO 364
368 A$(1) = STRING$(61, ".")
370 PRINT : PRINT "SETTING UP PLOTTING AREA " : PRINT
372 '
374 FOR I = 2 TO 40
376 A$(I) = "." + STRING$(59," ") + "."
378 NEXT I
380 '
382 A$(41) = A$(1)
384 '
386 'FIND MAX AND MIN X AND Y VALUES
388 '
390 XL = X(1) : XH = X(N) : YL=Y(1) : YH = Y(N)
392 '
394 FOR I = 1 TO N
396 IF X(I) < XL THEN XL = X(I) : GOTO 400
398 IF X(I) XH THEN XH = X(I)
400 IF Y(I) < YL THEN YL = Y(I) : GOTO 404
402 IF Y(I) YH THEN YH = Y(I)
404 NEXT I
406 '
408 ' CALCULATE X AND Y RANGES, SCALE DATA AND INSERT INTO STRING
ARRAY
410 '
412 XD = XH − XL
414 YD = YH − YL
416 '
418 FOR I = 1 TO N
420 X1 = INT(( 60 *( X(I) − XL)/ XD) + 0.5)
422 Y1 = 41 − INT(( 40 * (Y(I) − YL) / YD) +0.5)
424 P$ = A$(Y1)
426 X2 = X1 + 2
428 X3 = 64 − X2
430 T$ = MID$(P$,1,X1) + "o" + MID$(P$, X2, X3)
432 A$(Y1) = T$
434 NEXT I
436 '
438 INPUT "TITLE FOR YOUR GRAPH = "; T$
440 '
442 FOR I = 1 TO 5 : LPRINT : NEXT I
444 '
446 LPRINT T$ : LPRINT : LPRINT
448 LPRINT " Y MAX = "; YH
450 '
```

```
452 FOR I = 1 TO 41
454 LPRINT TAB(10) A$(I)
456 NEXT I
458 '
460 LPRINT " Y MIN = "; YL : LPRINT " X MIN = "; XL;
    TAB(60) "X MAX = "; XH
462 LPRINT " X AXIS SCALE : "; XD/60; " PER DOT TO DOT "; :
    LPRINT "INTERVAL"
464 LPRINT " Y AXIS SCALE : "; YD/40; " PER DOT TO DOT "; :
    LPRINT "INTERVAL"
466 LPRINT : LPRINT
468 '
470 ' PART D ::::::::::::::::::::::::::::::::::::::::::::::::::::::
472 '
474 SX = 0 : SY = 0 : XY = 0 : XX = 0 : YY = 0
476 '
478 FOR I = 1 TO N
480 SX = SX + X(I)
482 SY = SY + Y(I)
484 XY = XY + X(I) * Y(I)
486 XX = XX + X(I) * X(I)
488 YY = YY + Y(I) * Y(I)
490 NEXT I
492 '
494 ' PART E ::::::::::::::::::::::::::::::::::::::::::::::::::::::
496 '
498 XM = SX/N
500 YM = SY/N
502 SL = (XY − N * XM * YM) / (XX − N * XM ↑ 2)
504 IN = YM − SL * XM
506 R = (XY − N * XM * YM) / SQR( (XX − N * XM ↑ 2)*(YY − N*YM ↑ 2) )
508 DX = SQR( N/(N−1)) * SQR((XX/N) − (SX/N) ↑ 2)
510 DY = SQR( N/(N−1)) * SQR((YY/N) − (SY/N) ↑ 2)
512 SS = SQR(((DY/DX) ↑ 2) * ((1−R ↑ 2)/(N−2)))
514 SI = SQR(SS ↑ 2 * XX/N)
516 EY = DY * SQR(1 − R ↑ 2)
518 EX = DX * SQR(1 − R ↑ 2)
520 '
522 ' PART F ::::::::::::::::::::::::::::::::::::::::::::::::::::::
524 '
526 PRINT : PRINT STRING$(45,"*") : PRINT
528 PRINT " LINEAR REGRESSION OF Y ON X :" " : PRINT
530 SL = (INT(SL*10000 + 0.5))/10000
532 SS = (INT(SS * 10000 + 0.5)) / 10000
534 IN = (INT(IN*10000 + 0.5))/10000
536 SI = (INT(SI*10000 + 0.5))/10000
538 R = (INT(R*10000 + 0.5))/10000
540 PRINT "SLOPE = "; TAB(15) SL; TAB(25) "SD OF SLOPE = ";
TAB(45) SS
542 PRINT "INTERCEPT = "; TAB(15) IN; TAB(25) "SD OF INTERCEPT = ";
TAB(45) SI
544 PRINT TAB(10) "CORRELATION COEFFICIENT = "; TAB(40) R
546 PRINT
548 IF Q$= "N" THEN GOTO 558
550 LPRINT "SLOPE = "; TAB(15) SL; TAB(25) "SD OF SLOPE = ";
TAB(45) SS
552 LPRINT "INTERCEPT = "; TAB(15) IN; TAB(25)"SD OF INTERCEPT = ";
TAB(45) SI
554 LPRINT TAB(10) "CORRELATION COEFFICIENT = "; TAB(40) R
556 LPRINT
558 EY = (INT(EY*10000 +0.5))/10000
560 EX = (INT(EX*10000 + 0.5))/10000
562 PRINT "SE OF ESTIMATE OF Y FROM X = "; TAB(35) EY
```

```
564 PRINT : PRINT "SE OF ESTIMATE OF X FROM Y = "; TAB(35) EX;
"*********" : PRINT
566 '
568 IF Q$="N" THEN 574
570 LPRINT "SE OF ESTIMATE OF Y FROM X = "; TAB(35) EY
572 LPRINT : LPRINT "SE OF ESTIMATE OF X FROM Y = "; TAB(35) EX;
"******" : LPRINT
574 INPUT "PRESS RETURN TO CONTINUE "; C$
576 '
578 'PART G ::::::::::::::::::::::::::::::::::::::::::::::::::::::::
580 '
582 PRINT STRING$(45, "*")
584 PRINT "COMPARISON OF OBSERVED Y VALUES WITH THOSE CALCULATED ";
"FROM THE LINE OF BEST FIT" : PRINT
586 PRINT TAB(5) "OBS X"; TAB(15) "OBS Y"; TAB(25) "CALC Y";TAB(40)
"DIFF"; TAB(48) "DIFF/SE OF ESTIMATE"
588 '
590 FOR I = 1 TO N
592 C(I) = SL*X(I) + IN : C(I) = (INT(10000 * C(I) + 0.5))/10000
594 D(I) = Y(I) − C(I) : D(I) = (INT(10000 * D(I) + 0.5))/10000
596 T(I) = D(I)/EY : T(I) = (INT(100 * T(I) +0.5))/100
598 PRINT I; TAB(5) X(I); TAB(15) Y(I); TAB(25) C(I); TAB(40) D(I);
TAB(53) T(I);
600 IF ABS(T(I)) 1.96 THEN PRINT "*" ELSE PRINT " "
602 NEXT I
604 '
606 'PART H ::::::::::::::::::::::::::::::::::::::::::::::::::::::::
608 '
610 RV = 1 − R ↑ 2 : RV = (INT(RV * 10000 + 0.5))/100
612 PRINT : PRINT TAB(15) "RESIDUAL VARIANCE % = "; RV
614 IF RV <= 5 AND RV > 1 THEN PRINT TAB (15) "FIT IS AVERAGE" :
GOTO 626
616 IF RV <= 1 AND RV > 0.1 THEN PRINT TAB(15) "FIT IS GOOD" :
GOTO 626
618 IF RV <= 0.1 THEN PRINT TAB(15) "FIT IS EXCELLENT"
620 '
622 'PART I ::::::::::::::::::::::::::::::::::::::::::::::::::::::::
624 '
626 PRINT : INPUT "PRESS RETURN TO CONTINUE ";C$
628 PRINT : PRINT "DO YOU WISH TO DELETE ANY DATA POINTS AND "; :
INPUT "RECALCULATE THE REGRESSION ... Y/N "; Q$
630 IF Q$ = "N" THEN GOTO 666
632 INPUT "HOW MANY POINTS TO DELETE "; PD
634 '
636 FOR I = 1 TO PD
638 INPUT "INDEX NO. OF DATA PAIR TO BE DELETED "; OT
640 PRINT X(OT); Y(OT)
642 INPUT "IS THIS THE PAIR FOR DELETION ... Y/N "; Q$
644 IF Q$ = "N" GOTO 638
646 FOR J = (OT + 1) TO N
648 X(J−1) = X(J) : Y(J−1) = Y(J)
650 NEXT J
652 N = N − 1 : PRINT "EDITED : X OBS"; TAB(20) "Y OBS"
654 FOR K = 1 TO N : PRINT K; TAB(10) X(K); TAB(20) Y(K) : NEXT K
656 NEXT I
658 '
660 GOTO 358
662 '
664 ' PART J ::::::::::::::::::::::::::::::::::::::::::::::::::::::::
666 '
668 PRINT :PRINT : PRINT "DO YOU WISH TO CALCULATE VALUES OF X "; :
```

```
INPUT "FOR OBSERVED VALUES OF Y "; Q$
670 IF Q$ = "N" THEN GOTO 692
672 INPUT "HOW MANY UNKNOWNS TO CALCULATE "; UK
674 '
676 FOR I = 1 TO UK
678 INPUT "OBSERVED VALUE OF Y "; OY
680 CX = (OY − IN) / SL
682 CX = (INT(CX * 10000 + 0.5)) / 10000
684 PRINT "CALCULATED VALUE OF X = "; CX : PRINT
686 NEXT I
688 '
690 PRINT : PRINT : PRINT STRING$(45, "*")
692 STOP
```

1.2.2 Features of the linear regression program

Linear-regression analysis of y on x is the appropriate method for use with calibration data that are truly linear. Under these circumstances the concentrations in the standards are the independent variable, x, and the instrumental response are the dependent variable, y. This may appear to be an over-emphasis of the obvious, but unfortunately some instruments with built in calibration programs have been found to use linear-regression analysis of x on y. This has undoubtedly arisen because each value of x for a sample must be calculated from an observed instrumental response. However, the direct derivation of the equation

$$x = my + c$$

is an incorrect approach from the statistical point of view. We suggest that any instruments in the laboratory that have inbuilt linear-regression programs should be checked in order to find out if the correct approach has been adopted. Any potential instrument vendor should be questionned about the instrument's statistical procedures, particularly when the manufacturer has linearized the relationship between instrumental response and the concentration in the pre-packed standards, in accordance with an algebraic representation of the supposed chemical model of the reaction exploited by the instrument. In many instances, we believe that this approach intro-duces unnecessary complexity to the problem of curve fitting and can be misleading to some users who take the algebraic model as being strictly representational of what is actually occurring in their analyti-cal system. At best, it is a good approximation that has no advantage

when model-free techniques such as the cubic spline or a computerized version of the French Curve will yield equivalent results. In this
text, we have realized the computerized French Curve in a program
that uses the positive portion of a full sigmoid curve to fit data of
categories 2,3,4 and 5 (section 1.1).

PART C of the linear regression program offers a low resolution
plot of the calibration. In most instances this will not be required,
particularly if the analytical method is known to be highly linear. It
has been included, however, for those situations where duplicate (or
triplicate, etc.) readings have been made at each standard concentration, because the signal-to-noise ratio is poor, (e.g. in trace metal
analysis by atomic-absorption spectrophotometry), or because
reproducibility of the measurements is known to be poor for some
other reason, (e.g. the absorbance of the final test solution is
unstable). A printout of such a calibration graph can often permit the
analyst to make a more considered visual inspection of the quality of
the calibration data than he could on the screen. Outliers, particularly within each duplicate, triplicate ..., can often be noticed more
confidently and quickly from a graph than from a tabulated list of the
data. It should be noted that the graph plotting routine has a
resolution of only 1/40th of the span of the y data and 1/60th of the
span of the x data. These dimensions were chosen because on the
usual 80-column ten-pitch computer printer it produces a plot with
the conventional 2:3 height-to-width ratio. If this option is used
frequently and it is found that the scaling introduces unacceptable
discontinuities in either axis, then the dimension statement for A$ in
line 328 may be altered, along with lines 374, 376, 382, 420, 422, 428,
452, 462 and 464. An alternative to rescaling the program would be to
reselect the assigned values for the standards used in the assay, so that
lines 420 and 422 result in more appropriate positioning of the points
on the graph.

The equations used in Parts D and E have all been taken from
recognized statistical texts and were selected because they were
amenable to direct transcription into BASIC program statements.
We avoided reducing them to their simplest algebraic form because,
although this would have lead to shorter and perhaps more
"efficient" program code, it would have defeated our purpose of
providing easy to follow "self explanatory" programs that can be read
as part of this text. The standard deviations of the x and y data, lines
508 and 510 respectively, have had the Bessel's Correction Term,
$(n-1)/n$, included. This was an important inclusion because usually
there are relatively few data points in a calibration curve. Bessel's

Correction has the effect of increasing the estimate of the standard deviation of a data set when relatively few data items are involved in a statistical analysis.

The slope and intercept of a linear calibration graph provide important information about the current dynamic status of an assay. The intercept corresponds to the blank signal in the assay, so an increase in the intercept can often be indicative of deteriorating reagent quality. This can be important in trace metal analysis where, for example, the trace metal content of reagents used in sample preparation can constitute a significant proportion of the final total trace metal content. Extra high quality reagents such as the BDH Aristar range of chemicals may sometimes have to be used to reduce the blank. To assist the user to decide whether or not the intercept is significant, the program provides the standard deviation of the intercept. If the intercept divided by its standard deviation is greater than or equal to 1.96 then there is a high probability, $p <= 0.05$, that the intercept is significantly greater than zero. Such an observation may of course be irrelevant in assays where the lowest standard is far removed from zero and none of the samples in the batch have low concentrations. In this situation the "significant" intercept is the result of an extrapolation across a wide concentration span within which no data have been collected. Hence in non trace analysis the intercept assumes a largely notional importance and its only use is in the back calculation of concentrations in the samples.

The slope of the calibration line corresponds to the sensitivity of the assay and changes in the slope of the calibration line from batch to batch can signal deteriorating instrumental response, e.g. a photomultiplier in a spectrophotometer is approaching the end of its useful life or there has been some deterioration in reagents, e.g. labile reagents such as an immobilized enzyme or a radiolabelled protein. Clearly a thorough quality-control scheme should include batch-to-batch documentation of these two important assay parameters so that when the quality-control specimens in a batch fall outside the acceptance limits, the slope and intercept of the calibration can be re-examined with reference to the typical values obtained for successful batches.

The correlation coefficient, R, has been calculated without any further attempts to determine its significance. Where a calibration graph is linear, R must be highly significant for the assay to be useful. A Student's t could be calculated for the correlation coefficient but this is an inappropriate test for a set of calibration data. A Fischer's Z Test was considered for inclusion in the program because it provides a table-free method of calculating the confidence limits for R, but we felt that such information would be of interest to only a minority of users.

The standard errors of the estimates of y from x and x from y are the next items to appear on the printout and are produced by Part F. The standard error of the estimate of y from x is used in outlier detection and assessment schemes and is discussed in the next paragraph. The standard error of the estimate of x from a given value of y is probably best regarded as the "bottom line" parameter in the whole calibration graph fitting exercise. It informs the analyst that for this batch of samples the imprecision of each result that he reports will be equal to this standard deviation. It can be a sobering experience for a methods-development chemist to go back to the routine laboratory a few months after introducing his optimized method for determining "A" in "Z" and actually see what the standard deviation of x from y is when his method is subjected to day-to-day use. Comparison of this with the within batch and between-batch data he obtained in the more idealized environment of the development laboratory may lead him to revise his earlier expectations of the ruggedness of the new method.

The standard error of the estimate of y from x can be used to detect outliers. Part G presents a table of columns of figures titled OBSERVED X, OBSERVED Y, CALCULATED Y, DIFFER- ENCE (= observed y−calculated y), and DIFFerence/SE of the ESTIMATE. If any result in the last column is greater than or equal to 1.96 then there is a good chance ($p=0.95$) that the corresponding observed y value is an outlier. Such data points are automatically marked with an asterisk by the program. The user can opt to delete any such points in the next part of the program. Users should be aware that a value of 1.65 in this right hand column would be indicative of an outlier at the $p<=0.10$ level of significance and 1.44 an outlier at the $p<=0.15$ level of significance.

Part H of the program shows the residual variance expressed as a percentage, RV%, of the total variance of the y data, with a qualitative comment about the goodness of fit. The comments are selected according to the following criteria:

If RV%$\leqslant$5% and >1% then "Fit is Average"
If RV%$\leqslant$1% and > 0.1% then "Fit is Good"
If RV%$\leqslant$0.1% then "Fit is Excellent"

When the outlier data point deletion option is used in Part I, the effect on the subsequently recalculated line of best fit is usually most noticeable as an improvement in the Residual Variance, with only a

small or negligible improvement in the correlation coefficient. Note
that when the outlier data point deletion routine is used, one data pair
is deleted at a time. Consequently the data are renumbered after each
deletion and an updated table of the remaining data is presented,
(lines 646 to 656), before the user can enter the updated identification
number of the next data pair to be deleted.

1.2.3 Linear calibration program: worked examples

Table 1.1 lists a set of typical linear calibration data generated by
using the model equation:

$$Absorbance \times 1000 = 7.5 \times Concentration + 50$$

Table 1.1 — Data used to illustrate the performance of the linear
calibration program (symmetrical pseudo-random data).

Conc.	Absorbance Calculated from Model Equation × 1000	1 SD = Model Absorbance × 0.02 × 1000	SD multipliers as Selected from Table A1		Simulated Duplicate Absorbances × 1000	
1	50	1	+0.3,	−0.5	50,	51
10	125	3	+0.6,	−1.1	127,	122
20	200	4	+1.0,	−1.1	204,	196
30	275	6	+2.0,	−0.7	287,	271
40	350	7	+0.6,	−0.9	354,	344
50	425	9	+1.4,	−1.2	438,	414
60	500	10	+0.7,	−0.2	507,	498
70	575	12	+0.2,	−0.8	577,	565
80	650	13	+1.9	−1.2	675,	634
90	725	15	0 ,	−1.2	725,	707
100	800	16	+0.3,	−1.7	805,	773

The Gaussian Data Generator described in Appendix A was used,
with the assumption that the coefficient of variation of the absor-
bances was ±2%. The range of concentrations was set at 0–100 units
in steps of 10 units, with duplicate absorbance readings for each

standard. The standard deviations from Table A1 in the Appendix were set alternately at positive or negative; i.e. a symmetrical allocation of errors was generated for each absorbance duplicate reading.

When these 22 data points were processed by the linear calibration graph program it returned a slope of 7.42, an intercept of 52.8 and a residual variance of 0.2%. The data pairs numbered 17 and 22, (80,675 and 100,773), were marked with an asterisk indicating that they were possible outliers, and since the model equation predicted values of 80,650 and 100,800 then the program appeared to have made a reasonable identification. Note, however, that the program had not highlighted data-pair number 7, (30,287), which was +2.0 standard deviations from the target value of 275. Subsequent deletion of the two asterisked points resulted in the reporting of slope of 7.43, an intercept of 51.9 and a residual variance of 0.12%. Comparison of the two residual variances, 0.2% and 0.12%, illustrates our earlier comment that such a comparison provides a much clearer indication of an improvement in the fit to the remaining data than does the comparison of the two correlation coefficients, 0.9990 and 0.9994 respectively. Similarly, comparison of the standard error of the estimate of x from y indicated a substantial improvement from 1.47 to 1.06.

This first simulation was not entirely random in that the standard deviation multipliers were assigned alternately as positive or negative. In Table 1.2 the same data as in Table 1.1 were used, but the signs of the multipliers were chosen according to whether the random number in Table A2 was odd or even; if odd, the sign was assigned as negative.

The first pass of these data through the linear calibration program produced a slope of 7.57, an intercept of 50.3 and a residual variance of 0.18%. Data point number 20, (90,707), was marked with an asterisk and its subsequent exclusion reduced the residual variance to 0.12%. There were associated improvements in the standard error of the estimate of x from y; 1.36 dropped to 1.14 on the second pass. The corresponding values for the slope and intercept were 7.62 and 49.1 respectively. Table 1.3 summarizes the results from these two simulations.

The data in Table 1.3 show that the two best fits, if given an absorbance value of say 0.425 ($\times 1000$) from an unknown sample would return values of 50.2 and 49.3 respectively. These, and the predicted value, 50.0, are in agreement within the experimental error of this particular method.

Table 1.2 —Data used to illustrate the performance of the linear calibration program (fully random simulated data).

Concentration	1 SD = Model Absorbance × 0.02 × 1000	SD Multipliers as Selected from Table A1 with Signs Allocated Randomly		Simulated Duplicate Absorbances × 1000	
0	1	−0.3,	+0.5	50,	51
10	3	−0.6,	−1.1	123,	122
20	4	+1.0,	−1.1	204,	196
30	6	+2.0,	+0.7	287,	279
40	7	−0.6,	+0.9	346,	356
50	9	+1.4,	+1.2	438,	436
60	10	−0.7,	−0.2	493,	498
70	12	+0.2,	+0.8	577,	585
80	13	+1.9,	+1.2	675,	666
90	15	+0 ,	−1.2	725,	707
100	16	−0.3,	+1.7	795,	827

Table 1.3 — Results from the analysis of the data presented in tables 1.1 and 1.2.

	SLOPE	INTERCEPT	RV%	SE of Estimate x from y	No. of Data pairs
MODEL:	7.50	50.0	—	—	—
TABLE 1.1:	7.42	52.8	0.20	1.47	22
	7.43	51.9	0.12	1.06	20
TABLE 1.2:	7.57	50.3	0.18	1.36	22
	7.62	49.1	0.12	1.14	21

1.3   THE NON-LINEAR CALIBRATION GRAPH: FITTING BY CUBIC SPLINE

Analytical methods that produce calibration curves which approach exponential, hyperbolic, sigmoid or polynomial responses are not uncommon in analytical chemistry. They present the analytical chemist with the tedious task of preparing careful calibration graphs for each batch of analyses. It is therefore not surprising to find that wherever possible such analytical systems have had their responses linearized by some sort of mathematical fiddle. Often the latter has been arrived at by making some hypothesis as to the reaction mechanisms involved and then deriving a mathematical model that is amenable to linearization. At best these models are only approximations, and an analytical system that has been so mathematically modelled under ideal development laboratory conditions may not in fact adhere to that model when the usual variety of specimen types are presented to it for analysis in the routine testing laboratory. We have therefore chosen to avoid model-based methods of curve fitting and instead rely upon the smoothed cubic spline to fit our non-linear calibration graphs. This approach possesses considerable versatility, but it is not without hazards, as will become apparent when worked examples are examined.

A theoretical treatment of the cubic spline is beyond the scope of this text; references to some of the relevant literature are given in the bibliography. The cubic spline is the mathematical analogue of the engineer's spline, which consists of a long narrow flexible strip of wood or plastic which can be made to pass over or under a series of fixed pegs by bending it and/or loading it with strategically placed lead weights. This process can be expressed as a cubic of the form:

$$y = A_i + B_i x_i + C_i x_i^2 + D_i x_i^3$$

where x_i and y_i are the values at the ith peg.

For intermediate points between the pegs the equation takes the form :

$$y = A_i + B_i (x_j - x_i) + C_i (x_j - x_i)^2 + D_i (x_j - x_i)^3$$

where $x_i < x_j < x_{i+1}$.

In other words, the values of the coefficients A_i, B_i, C_i and D_i are entirely local and apply only along the interval of the curve between the ith and $(i+1)$th pegs. The cubic spline curve-fitting procedure is essentially a minimization procedure aimed at obtaining the values of these coefficients which give a smooth curve over the segment x_i to x_{i+1}, but with the added proviso that these must give a smooth progression from and to the curvatures in both the previous segment x_{i+1} to x_i and the following segment x_{i+1} to x_{i+2} respectively. This is achieved mathematically by ensuring that the first and second derivatives of the equation that applies at each x, y data point or **knot** (as it is referred to in spline fitting) are both equal. S. D. Conte and C. de Boor [1] have presented a concise description of this concept along with subroutines written in Fortran to achieve a cubic spline fit to non-linear data; (see Appendix B for the translations of these into BASIC), but these subroutines will not accomodate random noise in the data. The curve must pass and does pass through each data point. Reinsch [2] has described the theory behind a cubic spline fitting routine which involves a smoothing factor, or, more exactly, weighting factors for each point and a smoothing factor for the overall curve. Our BASIC program is based on an ALGOL program given by Reinsch [2], with the addition of data entry and output routines. For example, data points may be deleted after making the first trial fit, and an estimate of the goodness of fit of the spline to the experimental data is provided. The line numbering for this program was chosen to follow on directly from the program described in previous section, so that they can be selectable from a 'Menu' occupying program lines 1–299.

1.3.1   The SPLINE program
PART A
Lines 726–830 comprise the data array dimensioning and data entry routines. In analytical work, we frequently have duplicate or more readings at each standard concentration value; (x value). Cubic spline curve fitting, however, cannot handle these multiple readings (i.e. multiple knots at each x value) so the mean and standard deviation of the replicates have to be calculated for each standard, then these are passed to the cubic spline fitting routine. (Note the standard deviation is not required for the spline fitting but is essential information for the smoothing process.) For this reason two additional variables had to be introduced into our program, N, the total number of data pairs, line 730, and RP, the number of replicates of each standard, line 732. Reinsch used index numbers, N1 and N2, to index the data. In this program the value N1 is set to one, line 736, and N2 equated to (N/RP), line 732; i.e. N2 equals the number of standards in the assay. The replicate values for each standard are stored in the array YO(N2,RP) where YO is an abbreviation for 'y observed'. In lines 814 and 816 two further variables are introduced, LS and US, which correspond to the lower and upper limits of Reinsch's smoothing factor S.

$$LS = N2 + 1 - SQR(2*N2 + 2) : \text{line } 814$$
$$US = N2 + 1 + SQR(2*N2 + 2) : \text{line } 816$$

Reinsch recommended that S should initially be set equal to LS and modified by the user after an assessment of the first fit to the data. The weighting of each mean y value, calculated at line 804, is determined by the standard deviation of the y data routine, lines 790–810, for each standard concentration and its associated y values. Note that if there are only singlet y values, line 778 requests the user to input an estimate of the standard deviation of each y value. In summary, the data arrays for, say, five standards with triplicate readings of the y values are set up in our program according to the scheme:

X(1)	Y(1)	DY(1)	YO(1,1) → YO(1,3)
↓	↓	↓	↓
X(5)	Y(5)	DY(5)	YO(5,1) → YO(5,1)
Standard	Mean	SD of the	Triplicates of the
Concs.	Instrument	Triplicate	Instrument
	Response	Instrument	Readings
		Readings	

The user is given the opportunity to select a value of the smoothing factor, S, at lines 824–826. In practice a smoothing factor of zero ensures that the fitted curve passes through each point on the curve and a very high smoothing factor reduces the fit to a straight line. When the standard deviations of the data are small, (i.e., the range of the coefficients of variation are 2–5%) the recommended value of S (line 822) is usually too large, and a better fit can be obtained by using a smaller value than is recommended by the program. In contrast, when the data are associated with large coefficients of variation then a smoothing factor less than that recommended can result in quite unexpected oscillations between the knots. This is illustrated in one of the worked examples.

PARTS B TO G
These portions of the program correspond exactly in variable names and sequence to that used by Reinsch in his ALGOL program. The matrix of coefficients A_i, B_i, C_i and D_i are determined and stored in Part G, lines 960, 974, 962 and 972 respectively.

PART H
This section of the program, lines 984–1070, provides the user with a tabulated report on the degree of fit that has been obtained, and the opportunity to delete outliers from the data set before attempting a

new fit. The routine in lines 994–1016 determines the sum of squares of differences, SSD, between the mean y values in array Y(I) and the corresponding values calculated from the spline fit. By definition, the result of the spline fit at each X(I) datum point is equal to the corresponding value in the A(I) array, line 996. Line 1002 determines the sum of squares of differences between the observed y values and the corresponding spline-determined y value. Clearly, this is an overestimate because the spline has only been fitted to the mean y value, Y(I), at each datum point and not to all the y values in replicate, YO(I, 1 to RP). Its use, however, is justified because we require to identify outliers in the YO(I, 1 to RP) data that have an aberrant influence upon the mean Y(I) values. In line 1024 the sum of squares of differences is used to calculate a standard error of the estimate of y from x:

$$\text{SE of estimate of } y \text{ from } x = \text{SQR} \,[\text{SSD}/(N-4)]$$

Note that there are $N-4$ degrees of freedom because the statistics of polynomial regression require that the number of degrees of freedom equals the number of points fitted minus the number of parameters estimated [3] and we have estimated four (A, B, C and D) in the cubic equation.

In line 1042 the difference betwen each observed y value and the y value predicted by the spline is calculated and then divided by the standard error of the estimate (line 1044). In line 1046, if the absolute value of this calculation is greater than 1.96 then a possible outlier has been detected and is marked with a double asterisk. Outlier deletion is performed by Part I, which replaces user-selected values in the YO(I,RP) array by a marker value of -9999 that is detected by and dealt with as appropriate in lines 1002 and 1040.

Because this is a non-linear regression fitting routine, the conventional correlation coefficient is not applicable, and therefore cannot be used as an index of the goodness of fit. Instead we have used the reduced chi-squared test. This involves the solution of the equation:

$$\text{Reduced chi-squared} = \frac{1}{(N2-4)} \sum_{I=1}^{I=N2} \left(\frac{Y(I) - f(X(I))}{\text{SD of } Y(I)} \right)^2$$

where $f(X(I))$ is the value of the spline function for $x=X(I)$. The summation term is determined in line 1010 and the multiplication by the $1/(N2-4)$ term performed in line 1062. Note again that $(N2-4)$ is the number of degrees of freedom, because four parameters are estimated. Theoretically, a good fit is signified by a reduced chi-squared value of one, i.e. the fitted curve is on average within one standard deviation of the means of the y data. To determine the significance level of the reduced chi-squared result, the chi-squared distribution table is entered at the ONE DEGREE of freedom level. The table indicates that reduced chi-squared values greater than 3.84 or 6.64 are only encountered by chance alone in 5% and 1% of random trials respectively. These are the bases for the comments generated by lines 1066 and 1068. The chi-squared distribution table for one degree of freedom states that a chi-squared value of 0.46 coincides with the median of the distribution of chi, i.e. 50% of chi values will be greater than this. A chi-squared value of 1.07 occurs at a point in the distribution where only 30% of random trials will exceed this value. This implies that 70% of the reduced chi-squared values will be less than 1.07, but because of the asymmetrical and non-gaussian shape of the chi-squared distribution for one degree of freedom, confidence limits about this point in the distribution cannot be easily derived or visualized. Note that whenever the smoothing factor is set equal to zero, the result of the reduced chi squared will always be zero!

PART I
This portion of the program is concerned with the removal of candidate outliers as directed by the user. It caters for two situations. Lines 1086–1104 replace outliers with a marker value of -9999 in the situation where replicate values of y are involved. Lines 1114–1134 cater for the situation where only singlet y observations have been made, in which case deletion of a point requires only that the data arrays be repacked, lines 1122–1126, and the counter N2 decremented (line 1130).

The replicate y observations require that the values of $DY(I)$ be recalculated wherever a deletion is made, (lines 1146–1152 and lines 1198–1200). In addition, if say two y observations out of a triplet have been deleted, then the user must enter an appropriate value for the standard deviation of the remaining point, (lines 1188–1192).

When the user-directed deletions have been made, the program loops back to line 814 in Part A so that a new value for the smoothing factor, S, can be selected.

PART J
In the discussion of Part A of this program, it was mentioned that certain selections of the smoothing factor, S, can be associated with unexpected oscillations. These are not apparent from the data

presented to the user by Part H, because they occur between datum points. We consider it essential that a graphical presentation of the observed x, y data and the fitted curve must be available to the user for visual inspection. Part J of the program provides both a scrolling plot on the monitor and a hard copy on an 80 column printer. To keep the programming relatively straightforward, the plots on both the monitor and the printer are presented with the y axis HORIZON-TAL, and the intersection of the x and y axes is located at the top left hand corner of the monitor. The y axis is clearly labelled on the hard copy printout, (line 1218). The object is to present a plot to allow an assessment of whether or not oscillations and/or a biased fit have been obtained, requiring use of a new smoothing factor (line 1306). The dimensions of the plot are 60 units along the y axis and 40 units along the x axis. As each line of the plot is produced, the loop in lines 1246–1250 checks to see if the current x-value coincides with a value in the $X(I)$ array. If it does, then the x axis is marked with an 'S' to indicate the position of a standard, line 1276, and a 'o' is plotted at the appropriate position for the corresponding $Y(I)$ value so as to distinguish it from the interpolated y values plotted as "∗". In lines 1272 and 1278 there are delay loops:

$$\text{FOR T} = 1 \text{ TO } 300 : \text{NEXT T}$$

to slow down the plotting process so that the user can assess the spline fit on the monitor. The plot routine also tests to determine whether an interpolated point is off scale, i.e. less than the minimum y value, MINY; line 1268. If it is, then 'OS' for 'off scale' is printed at Tab 0.

Part J concludes with a routine (lines 1312–1356), for the calcula-tion of x values from observed instrument responses obtained from measurements on samples. For a given y value the routine 1322–1328 determines which standards bracket the unknown. The Newton-Raphson minimization method is then used in lines 1338–1348 to determine the corresponding value of x to within 0.1% (line 1344).

1.3.2   Listing of SPLINE

```
698 'SPLINE.BAS
700 'THE NONLINEAR CALIBRATION CURVE
702 'PROGRAM USING THE REINSCH SMOOTHING SPLINE
704 '
706 'T F HARTLEY
708 'COMPUTERIZED QUALITY CONTROL
710 'PUBLISHED BY ELLIS HORWOOD, ENGLAND, 1986
712 ' CLEAR SOME SCREEN
714 FOR I = 1 TO 20 : PRINT : NEXT I
716 PRINT STRING$(70, "*")
```

718 PRINT : PRINT "THE NON"LINEAR CALIBRATION CURVE :"
720 PRINT "LINE OF BEST FIT USING THE SMOOTHING SPLINE"
722 PRINT : PRINT STRING$(70, "*"): PRINT
724 '
726 'PART A ::
728 '
730 INPUT "HOW MANY CALIBRATION X,Y DATA PAIRS DO YOU HAVE ";N:
    PRINT
732 INPUT "HOW MANY REPLICATES ARE THERE FOR EACH VALUE OF X "; RP
    : N2 = N/RP : PRINT
734 '
736 N1 = 1 : M1 = 0 : M2 = N2+1
738 DIM X(N2), Y(N2), DY(N2), YO(N2,RP)
740 DIM R(M2), R1(M2), R2(M2), T(M2), T1(M2), U(M2), V(M2), A(M2),
    B(M2), C(M2), D(M2)
742 R(M1)=0 : R(N1)=0 : R1(N2)=0 : R2(N2)=0 : R2(M2)=0 : U(M1)=0 :
    U(N1)=0 : U(N2)=0 : U(M2)=0
744 Q$ = ""
746 '
748 FOR I = 1 TO N2
750 PRINT I; " "; : INPUT " X = "; X(I)
752 IF I 1 AND X(I) < X(I"1) THEN PRINT "* X'S MUST BE ENTERED IN
    AN INCREASING SEQUENCE *" : PRINT : CLEAR : GOTO 722
754 '
756 FOR J = 1 TO RP
758 PRINT TAB(15); : INPUT "Y = "; Y
760 YO(I,J) = Y
762 IF RP = 1 THEN Y(I) = Y
764 NEXT J
766 '
768 PRINT TAB(25) "ALL DATA OK Y/N ...... "
770 Q$ = INKEY$: IF Q$ = "Y" THEN GOTO 774
772 IF Q$ = "N" THEN GOTO 750 ELSE 770
774 IF RP = 1 THEN GOTO 778
776 GOTO 780
778 INPUT "YOUR ESTIMATE OF THE SD OF THIS Y "; DY(I)
780 NEXT I
782 '
784 IF RP = 1 THEN GOTO 814
786 SY = 0 : YY = 0 : DI = 0
788 '
790 FOR I = 1 TO N2
792 '
794 FOR J = 1 TO RP
796 SY = SY + YO(I,J)
798 YY = YY + (YO(I,J)) ↑ 2
800 NEXT J
802 '
804 Y(I) = SY / RP
806 DY(I) = SQR((YY/ RP) − (SY / RP) ↑ 2)
808 SY = 0 : YY = 0 : DI = 0
810 NEXT I
812 '
814 LS = N2+1 − SQR(2*N2 + 2)
816 US = N2+1 + SQR(2*N2 + 2)
818 S = LS
820 PRINT : PRINT "SMOOTHING FACTOR RANGE = ";LS;" − ";US
822 PRINT : PRINT "SMOOTHING FACTOR RECOMMENDED = ";LS : PRINT
824 INPUT "IS THIS ACCEPTABLE ";Q$
826 IF Q$="N" THEN INPUT "YOUR CHOICE OF SMOOTHING FACTOR = ";S
828 Q$=""
830 '

```
832 'PART B ::::::::::::::::::::::::::::::::::::::::::::::::::::
834 '
836 P = 0 : M1 = N1 + 1 : M2 = N2 − 1
838 H = X(M1) − X(N1) : F = ( Y(M1) − Y(N1) )/H
840 '
842 FOR I = M1 TO M2
844 G = H : H = X(I+1) − X(I)
846 E = F : F = ( Y(I+1) − Y(I) )/H
848 A(I) = F − E : T(I) = 2 * (G + H)/3 : T1(I) = H/3
850 R2(I) = DY(I−1)/G : R(I) = DY(I+1)/H
852 R1(I) = − DY(I)/G − DY(I)/H
854 NEXT I
856 '
858 'PART C ::::::::::::::::::::::::::::::::::::::::::::::::::::
860 '
862 FOR I = M1 TO M2
864 B(I) = R(I) ↑ 2 + R1(I) ↑ 2 + R2(I) ↑ 2
866 C(I) = R(I) * R1(I+1) + R1(I) * R2(I+1)
868 D(I) = R(I) * R2(I+2)
870 NEXT I
872 F2 = −S
874 '
876 'PART D :::::: NEXT ITERATION :::::::::::::::::::::::::::::::
878 '
880 FOR I = M1 TO M2
882 R1(I−1) = F*R(I−1) : R2(I−2) = G*R(I−2)
884 R(I) = 1/(P* B(I) + T(I) − F * R1(I−1) − G * R2(I−2) )
886 U(I) = A(I) − R1(I−1)*U(I−1) − R2*(I−2)*U(I−2)
888 F = P*C(I) + T1(I) − H*R1(I−1) : G = H : H = D(I) * P
890 NEXT I
892 '
894 'PART E ::::::::::::::::::::::::::::::::::::::::::::::::::::
896 '
898 FOR I = M2 TO M1 STEP −1
900 U(I) = R(I) * U(I) − R1(I) * U(I+1) − R2(I) * U(I+2)
902 NEXT I
904 '
906 E = 0 : H = 0
908 '
910 FOR I = N1 TO M2
912 G = H : H = ( U(I+1) − U(I) ) / (X(I+1) − X(I))
914 V(I) = (H−G) * DY(I) ↑ 2 : E = E + V(I)*(H−G)
916 NEXT I
918 '
920 'PART F ::::::::::::::::::::::::::::::::::::::::::::::::::::
922 '
924 V(N2) = − H*DY(N2) ↑ 2 : G = V(N2) : E = E − G*H
926 G = F2 : F2 = E*P ↑ 2
928 IF F2 >= S OR F2 <= G THEN GOTO 954
930 '
932 F = 0 : H = (V(M1) − V(N1)) / (X(M1) − X(N1))
934 '
936 FOR I = M1 TO M2
938 G = H : H = (V(I+1) − V(I))/ (X(I+1) − X(I))
940 G = H − G − R1(I−1)* R(I−1) − R2(I−2) * R(I−2)
942 F = F + R(I)*G ↑ 2 : R(I) = G
944 NEXT I
946 H = E − P*F
948 IF H <= 0 THEN GOTO 954
950 P = P + (S − F2) / (( SQR(S/E) + P) * H) : GOTO 876
952 '
954 'PART G ::::::::::::::::::::FIN::::::::::::::::::::::::::::::
```

```
956 '
958 FOR I = N1 TO N2
960 A(I) = Y(I) − P*V(I)
962 C(I) = U(I)
964 NEXT I
966 '
968 FOR I = N1 TO M2
970 H = X(I+1) − X(I)
972 D(I) = ( C(I+1) − C(I)) /(3*H)
974 B(I) = ( A(I+1) − A(I))/H − (H*D(I) + C(I) ) * H
976 NEXT I
978 '
980 ' END SMOOTH
982 '
984 'PART H ::::::::::::::::::::::::::::::::::::::::::::::::::::::::
986 '
988 PRINT : PRINT STRING$(70, "*") : PRINT : PRINT "TABLE OF "; :
    PRINT "CALCULATED Y FROM CUBIC SPLINE FIT vs OBSERVED Y's" :
    PRINT
990 CHI = 0 : SSD = 0 : L = 0
992 '
994 FOR I = 1 TO N2
996 Y = A(I)
998 '
1000 FOR J = 1 TO RP
1002 IF YO(I,J) = −9999 THEN GOTO 1006
1004 SSD = SSD + (YO(I,J) − Y) ↑ 2
1006 NEXT J
1008 '
1010 CHI = CHI + ( (Y(I) − Y) / DY(I) ) ↑ 2
1012 '
1014 '
1016 NEXT I
1018 '
1020 PRINT TAB(5) "OBS X"; TAB(15) "OBS Y"; TAB(25) "CALC Y";TAB(40)
     "DIFF"; TAB(48) "DIFF/SE OF ESTIMATE" : PRINT
1022 T = 0
1024 EY = SQR( SSD/(N−4) ) : EY = (INT( 10000*EY + 0.5)) / 10000
1026 '
1028 FOR I = 1 TO N2
1030 Y = A(I)
1032 Y = ( INT(10000*Y + 0.5)) / 10000
1034 '
1036 FOR J = 1 TO RP
1038 T = T + 1
1040 IF YO(I,J) = −9999 THEN PRINT T ; " ! THIS POINT WAS DELETED":
     GOTO 1048
1042 D = YO(I,J) − Y
1044 PRINT T; TAB(5) X(I); TAB(15) YO(I,J); TAB(25) Y; TAB(40)
     (INT(D*100 + 0.5)) / 100; TAB(53) (INT(D / EY*100 + 0.5))/100;
1046 IF ABS(D/EY) > 1.96 THEN PRINT " **" ELSE PRINT
1048 L = L + 1 : IF L = 20 THEN L=0 : PRINT :
     PRINT "NOTE INDEX NUMBER OF CANDIDATE OUTLIERS IN THIS LIST":
     INPUT "PRESS RETURN TO CONTINUE LIST...."; Q$
1050 NEXT J
1052 '
1054 NEXT I
1056 '
1058 PRINT
1060 PRINT "STD ERROR OF THE ESTIMATE OF Y FROM X = "; EY
1062 CHI = CHI / (N2 − 4)
```

```
1064 PRINT "REDUCED CHI SQUARED = "; ( INT(10000 * CHI + 0.5)) /
      10000; "Target is 1.0 ";
1066 IF CHI >= 3.84 THEN PRINT " FIT IS POOR" ELSE PRINT
1068 IF CHI >= 6.64 THEN PRINT " FIT IS VERY POOR"
1070 '
1072 ' PART I ::::::::::::::::::::::::::::::::::::::::::::::::::
1074 '
1076 PRINT : PRINT "DO YOU WISH TO DELETE ANY DATA POINTS AND "; :
      INPUT "RECALCULATE THE FIT .. Y/N"; Q$
1078 IF Q$ = "N" THEN GOTO 1210
1080 INPUT "HOW MANY POINTS DO YOU WISH TO DELETE ";PD
1082 IF RP = 1 GOTO 1114
1084 '
1086 FOR I = 1 TO PD
1088 INPUT "INDEX NUMBER OF DATA PAIR TO BE DELETED "; OT
1090 OT = OT − 1
1092 K = 1 + INT( OT / RP )
1094 L = 1 + RP * ( (OT/RP) − INT( OT/RP) )
1096 PRINT X(K); YO(K,L)
1098 INPUT "IS THIS THE PAIR FOR DELETION.... Y/N "; Q$
1100 IF Q$ = "N" THEN GOTO 1088
1102 YO(K,L) = −9999
1104 NEXT I
1106 '
1108 GOTO 1138 : 'AMEND THE DATA ARRAYS X(), Y(), YO()
1110 '
1112 'SINGLETS − NO REPLICATES
1114 FOR I = 1 TO PD
1116 INPUT "INDEX NUMBER OF THE PAIR TO BE DELETED "; OT
1118 PRINT : PRINT X(OT);" ";Y(OT); " " ; :
      INPUT "IS THIS THE RIGHT PAIR "; Q$ : IF Q$ = "N" THEN GOTO
      1116
1120 '
1122 FOR J = (OT + 1) TO N2
1124 X(J−1) = X(J) : Y(J−1) = Y(J) : DY(J−1) = DY(J) : YO(J−1,1) =
      YO(J,1)
1126 NEXT J
1128 '
1130 N2 = N2 − 1
1132 NEXT I
1134 GOTO 814
1136 'END OF SINGLET DELETIONS
1138 DI = 0 : SY = 0 : YY = 0
1140 FOR I = 1 TO N2
1142 'EXAMINE THE DATA ARRAYS FOR −9999 AND AMEND &/OR REPACK
1144 '
1146 FOR J = 1 TO RP
1148 IF YO(I,J) = −9999 THEN DI = DI + 1 : GOTO 1152
1150 SY = SY + YO(I,J) : YY = YY + ( YO(I,J) ) ↑ 2
1152 NEXT J
1154 IF DI = RP THEN GOTO 1156 ELSE 1182
1156 'REPACK DATA ROUTINE
1158 FOR K = I TO N2 − 1
1160 X(K) = X(K+1)
1162 Y(K) = Y(K+1)
1164 '
1166 FOR L = 1 TO RP
1168 YO(K,L) = YO(K+1,L)
1170 NEXT L
1172 '
1174 NEXT K
1176 '
```

```
1178 N2 = N2 − 1 : GOTO 1202
1180 '
1182 IF RP − DI = 1 THEN GOTO 1184 ELSE 1198
1184 'SD OF REMAINING POINT
1186 '
1188 FOR K = 1 TO RP
1190 IF YO(I,K) <> −9999 THEN PRINT "REMAINING Y VALUE = "; YO(I,K)
      : INPUT "YOUR ESTIMATE OF THE SD OF THIS Y "; DY(I)
      : Y(I) = YO(I,K)
1192 NEXT K
1194 GOTO 1202
1196 '
1198 Y(I) = SY / (RP − DI)
1200 DY(I) = SQR(( YY / (RP − DI)) − ( SY / (RP − DI)) ↑ 2 )
1202 SY = 0 : YY = 0 : DI = 0
1204 NEXT I
1206 GOTO 814
1208 '
1210 'PART J ::::::::::::::::::::::::::::::::::::::::::::::::::::::
1212 '
1214 PRINT "PREPARE THE PRINTER FOR GRAPH PLOTTING ... THEN PRESS" :
      INPUT " RETURN ..";Q$
1216 INPUT "TITLE FOR YOUR GRAPH "; T$ : LPRINT : LPRINT : LPRINT :
      LPRINT T$ : LPRINT
1218 LPRINT TAB(20) − − − − − Y A X I S − − − − − >"
1220 LPRINT : LPRINT TAB(5) " "; STRING$(61, "|")
1222 PRINT TAB(5) " "; STRING$(61, "|")
1224 MINY = Y(1) : MAXY = Y(N2)
1226 '
1228 FOR I = 1 TO N2
1230 IF Y(I) < MINY THEN MINY = Y(I) : GOTO 1234
1232 IF Y(I) > MAXY THEN MAXY = Y(I)
1234 NEXT I
1236 YI = (MAXY − MINY) / 60 : XI = ( X(N2) − X(1) ) / 40 :
      OS = MINY
1238 '
1240 FOR J = 0 TO 40
1242 X = X(1) + J * XI
1244 '
1246 FOR L = 1 TO N2
1248 IF ABS(X − X(L)) < XI/2 THEN GOTO 1274
1250 NEXT L
1252 '
1254 IF X = X(N2) THEN GOTO 1264
1256 FOR K = 2 TO N2
1258 IF X >= X(K−1) AND X < X(K) THEN R = X − X(K−1) :
      Y = INT(( A(K−1) + B(K−1)*R + C(K−1)*R ↑ 2 + D(K−1)*R ↑ 3 − OS)
      / YI + 0.5 ) : GOTO 1268
1260 NEXT K
1262 '
1264 Y = INT(( A(N2) − OS )/YI + 0.5)
1266 '
1268 IF Y < 0 THEN PRINT "OS"; TAB(5) "−" : LPRINT "OS";TAB(5)"−":
      GOTO 1282
1270 LPRINT TAB(5) "−"; TAB(Y + 6) "*"
1272 PRINT TAB(5) "−"; TAB(Y + 6) "*"
      : FOR T = 1 TO 300 : NEXT T : GOTO 1282
1274 Y = INT( ( Y(L) − OS ) / YI + 0.5 )
1276 LPRINT TAB(5) "S"; TAB( Y + 6 ) "o"
1278 PRINT TAB(5) "S"; TAB(Y+6) "o" : FOR T = 1 TO 300 : NEXT T
1280 '
1282 NEXT J
```

```
1284 '
1286 LPRINT : LPRINT : LPRINT
1288 LPRINT "Y AXIS MINIMUM = ";OS;" Y INTERVALS = "; YI;
     " Y AXIS MAXIMUM = "; MAXY
1290 LPRINT : LPRINT "X AXIS MINIMUM = "; X(1);" X INTERVALS = ";
     XI ; " X AXIS MAXIMUM = "; X(N2)
1292 '
1294 LPRINT : LPRINT "SMOOTHING FACTOR = ";S : LPRINT
1296 '
1298 LPRINT : LPRINT "REDUCED CHI SQUARED = ";CHI;
     " S E OF ESTIMATE OF Y FROM X = ";EY
1300 IF CHI >= 3.84 THEN LPRINT "FIT IS POOR" :LPRINT :LPRINT :LPRINT
1302 IF CHI >= 6.64 THEN LPRINT "FIT IS VERY POOR" : LPRINT :
     LPRINT : LPRINT
1304 '
1306 PRINT : PRINT "DO YOU WISH TO TRY ANOTHER SMOOTHING FACTOR "; :
     INPUT "...Y/N"; Q$
1308 IF Q$ = "Y" THEN GOTO 818
1310 PRINT : PRINT STRING$(70, "*") : PRINT
1312 INPUT "HOW MANY UNKNOWN X'S DO YOU HAVE TO CALCULATE ";UK
1314 '
1316 FOR I = 1 TO UK
1318 PRINT : PRINT "NO. ";I; : INPUT " Y = "; P
1320 '
1322 FOR K = 1 TO N2 − 1
1324 IF P >= Y(K) AND P <= Y(K+1) THEN GOTO 1334
1326 IF Y(1) > Y(N2) AND P = Y(K+1) AND P <= Y(K) THEN GOTO 1334
1328 NEXT K
1330 '
1332 PRINT "THIS Y VALUE IS OUTSIDE THE LIMITS OF THE CALIBRATION"; :
     PRINT " CURVE" : GOTO 1356
1334 Z = (X(K+1) − X(K)) / 2
1336 '
1338 FOR A = 1 TO 100
1340 R5 = B(K) + 2*C(K)*Z + 3*D(K)*Z ↑ 2
1342 R6 = A(K) " P + B(K)*Z + C(K)*Z ↑ 2 + D(K)*Z ↑ 3
1344 IF ABS(R6/R5) < 0.001*Z THEN GOTO 1352
1346 Z = Z − R6/R5
1348 NEXT A
1350 '
1352 X = X(K) + Z
1354 PRINT " CALCULATED X = "; (INT(10000*X +0.5))/
     10000 : PRINT
1356 NEXT I
1358 '
1360 STOP
```

1.3.3   Cubic SPLINE fitting program : worked examples

Table 1.4 lists a set of duplicates selected such that the mean of each fits the hyperbolic equation:

$$y = 100 \, (1 - e^{-0.4x})$$

A computer plot of these data, along with the cubic spline line of best

Table 1.4 — Hyperbolic data used to assess the performance of the smoothed cubic spline curve-fitting program.

x	Calculated y	'Observed' Duplicate y-values	
1	32.9	31.9	33.9
2	55.1	54.1	56.1
3	69.9	68.9	70.9
4	79.8	78.8	80.8
5	86.5	85.5	87.5
6	90.9	89.9	91.9
7	93.9	92.9	94.9
8	95.9	94.9	96.9
10	98.2	97.2	99.2
14	99.6	98.6	100.6

fit calculated with the smoothing factor equal to zero is shown in Fig. 1.1. There are ten values of x, so the program commenced processing with a minimum Reinsch smoothing factor of $10+1-SQR(2\times 10+2)=6.31$, with the result that the reduced chi-squared of the fit was 1.05 and the SE of the estimate of y from x was 1.43. The latter value revealed that the fit at the first point, $x=1$ and $y=31.9$, was poor; the calculated y value is 34.83 and the difference divided by the SE of the estimate exceeded the ±1.96 limit ($p=0.05$). This was not surprising because it occurred at the beginning of the curve, so the fitting process had been provided with no information on curvature of any 'previous' segment.

A more inforative evaluation of the goodness of fit was obtained by examining the results of back calculation of x values for given values of y; $x=[\ln(1-y/100)]/(-0.4)$. Table 1.5 presents the results of such calculations for y values located at strategic points along the curve.

These data suggested that the fit was tolerable over most of the curve but was definitely unacceptable for values of y greater than 95. The simulation data in Table 1.4 consisted of such 'well behaved' duplicates that it was reasonable to rerun the fit using a smaller smoothing factor. A smoothing factor of zero, for example, would ensure that the fit passed through each of the means of the duplicates (which are the exact solutions to the test equation). When this was done, no data points were marked as possile outliers and the results of

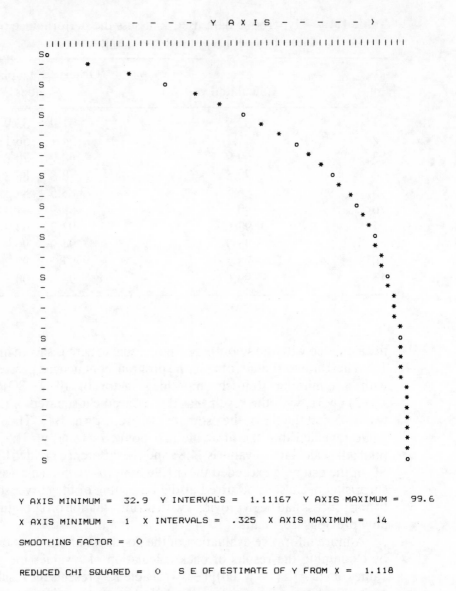

Fig. 1.1 — Computer-generated plot of the hyperbolic data in Table 1.4, processed
by the cubic spline program with a smoothing factor equal of zero.

the back-calculation of x for the same observed y values as shown in
Table 1.5 were considerably more satisfactory (Table 1.6). The SE of
the estimate of y from x was also reduced to 1.12.

A sigmoid curve is not unusual in analytical chemistry particularly
where enzyme or competitive protein-binding assays are involved.
The smoothed cubic spline can also be used to fit such data with

Table 1.5 —Theoretical and calculated values of x for the hyperbolic curve (smoothing factor = 6.31.

Observed y	Theoretical x	x calculated from the cubic spline fit	
45.1	1.500	1.530	(+2.0%)
75.3	3.496	3.530	(+1.0%)
88.9	5.496	5.450	(−0.8%)
95.0	7.489	7.407	(−1.1%)
99.2	12.07	11.43	(−6.0%)

Table 1.6 — Theoretical and calculated values of x for the hyperbolic curve (smoothing factor = 0).

Observed y	Theoretical x	x calculated from the cubic spline fit	
45.1	1.500	1.521	(+1.4%)
75.3	3.496	3.500	(+0.1%)
88.9	5.496	5.494	(0%)
95.0	7.489	7.500	(0%)
99.2	12.07	11.99	(−0.7%)

acceptable results. Table 1.7 presents some simulated calibration data based on the equation:

$$y = 50 \left(\frac{1}{1 + \text{EXP}[-0.01(x - 400)]} \right)$$

Once again, points were artificially chosen so that they were excellent duplicates on either side of the true mean. The computer plot of the data in Table 1.7 is shown in Fig. 1.2 along with the cubic spline line of best fit, calculated for smoothing factor zero.

The starting smoothing factor was 7.10, which resulted in a reduced chi-squared of 0.99 and an SE of the estimate of y from x of 1.29. The latter revealed that the fit in the vicinity of the duplicate points at $x=400$ was poor. The fit passed very close to the point 400,

Table 1.7 — Sigmoid data used to assess the performance of the smoothed cubic spline curve-fitting program.

x	Calculated y	'Observed' duplicate y-values	
0	0.899	0.80	1.00
80	1.96	1.86	2.06
160	4.16	4.06	4.26
240	8.40	8.300	8.50
320	15.5	15.4	15.6
400	25.0	24.0	26.0
480	34.5	33.5	36.5
560	41.6	40.6	42.6
640	45.8	44.8	46.8
720	48.0	47.0	49.0
800	49.1	48.0	50.0

33.5 and consequently was significantly removed from the point 400, 36.5. Five y values were used to back-calculate the corresponding values of x, and the results are shown in Table 1.8.

Overall, the results were not consistently too high or too low and three of the deviations were unacceptable. For comparison, the same sigmoid curve data were fitted with a smoothing factor of zero, and the results are presented in Table 1.9. The fit was considerably improved, with only the first estimate being unacceptable. Again this observed y of 1.33 fell within the unreliable first segment of the cubic spline fit. The SE of the estimate of y from x was 0.96; i.e. less than for the fit using the recommended smoothing factor.

These two examples have been based on good duplicates and adequate numbers of data points, which have permitted confident description of the known curve. Real assays often lack such adequate data, and the next is a typical example. The data were taken from an IgE protein radioimmunoassay method used in our laboratory some years ago. They illustrate how the recommended smoothing factor provided a better fit through the data points than a zero factor. The latter, in fact, resulted in the bizarre oscillations between the data points shown in Fig. 1.3. Without the plots of the fit on the screen and the printer, the user would have been unaware of these oscillations, and on the basis of the tolerable fit, (SE of the estimate of y from

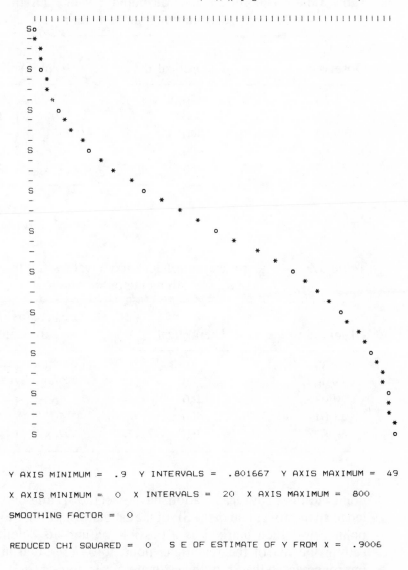

Fig. 1.2—Computer-generated plot of the sigmoid data (Table 1.7) processed by the
cubic spline program with a smoothing factor of zero.

$x=228$), to the rather poor data might have continued to use this fit to
calculate the concentrations in the unknowns. In this example, a
much improved fit was obtained when a smoothing factor of 4.76 was
used, as shown in Fig. 1.4 and Table 1.10. Nevertheless, point 4 was
marked as a possible outlier, and the SE of the estimate of y from
$x=367$, was worse than for the zero smoothing factor. Deletion of

Table 1.8 — Theoretical and calculated x values for the sigmoid curve, smoothing factor = 7.10.

Observed y	Theoretical x	x calculated from the cubic spline fit	
1.33	40.0	38.9	(−2.8%)
5.96	200	199	(−0.5%)
29.93	440	451	(+2.5%)
44.04	600	626	(+4.3%)
48.67	760	753	(−0.9%)

Table 1.9 — Theoretical and calculated x values for the sigmoid curve, smoothing factor = zero.

Observed y	Theoretical x	x calculated from the cubic spline fit	
1.33	40.0	38.1	(−4.8%)
5.96	200	200	(0%)
29.93	440	60	(−1.0%)
44.04	600	602	(+0.3%)
48.67	760	768	(+1.1%)

point 4 followed by recalculation, having set the SD for point 3 to 375, led to an improvement in the SE of the estimate of y from x to 273, but point 2 was then marked as a possible outlier. Its deletion, with setting the SD of the remaining point to 337 resulted in a small improvement in the SE of the estimate of y from x, (=264), but then point 3 was marked as a possible outlier. The program was clearly highlighting the inadequacies of the original data in the early portion of the calibration curve, and there was no advantage in proceeding with the process of eliminating any more points.

This example highlights the importance of the user's judgement of the validity of the curve-fitting process. Data points must only be removed from the computation after careful consideration, and under normal analytical circumstances the deletion of more than one possible outlier should not be permitted.

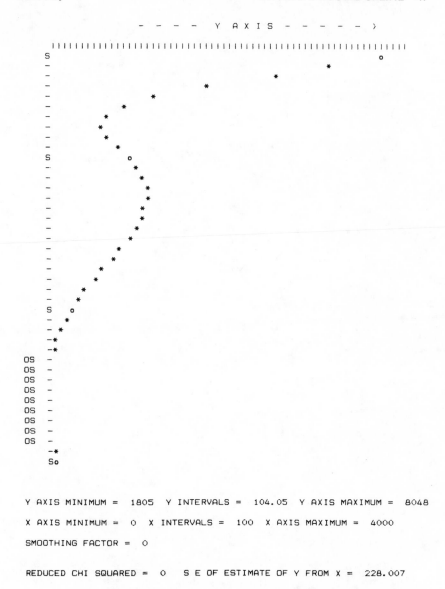

Fig. 1.3 — Computer-generated plot of the IgE assay data from Table 1.10, processed by the cubic spline program (smoothing factor =0).

The next example is based on thyroxine RIA data [5]. The original paper just mentioned the cubic spline, but concentrated on 5 other methods of fitting:

A: Manual plotting with use of a Flexicurve.
B: Unweighted linear regression with use of the logit–log [6].

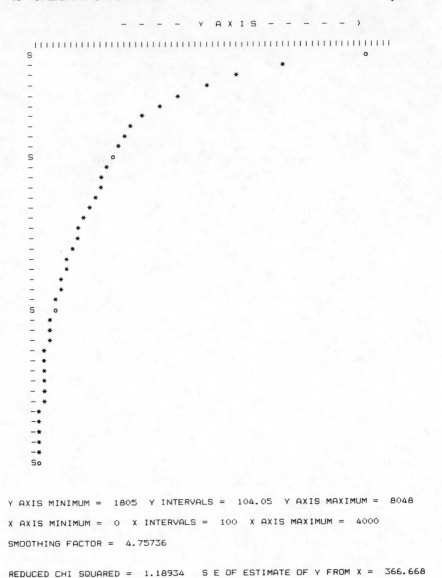

Y AXIS MINIMUM =   1805   Y INTERVALS =   104.05   Y AXIS MAXIMUM =   8048

X AXIS MINIMUM =   0   X INTERVALS =   100   X AXIS MAXIMUM =   4000

SMOOTHING FACTOR =   4.75736

REDUCED CHI SQUARED =   1.18934   S E OF ESTIMATE OF Y FROM X =   366.668

Fig. 1.4 — Computer-generated plot of the IgE assay data from Table 1.10, processed by the cubic spline program (smoothing factor = 4.76).

C: Four-parameter logistic model [7].
D: Four-parameter logistic model [8].
E: Amersham simplified model [9].

The data are listed in Table 1.11 and plotted in Fig. 1.5 along with the cubic spline for a smoothing factor of 6.31. Duplicate counts for a

Table 1.10 — Data from the IgE assay, along with the results of the cubic spline fit (smoothing factor = 4.76).

#	OBS x	OBS y	CALC y	DIFF	DIFF/SE OF ESTIMATE
1	0	7355	7550.76	−195.76	−0.53
2	0	7832	7550.76	281.24	0.77
3	10	7699	7372.82	326.18	0.89
4	10	8397	7372.82	1024.18	2.79**
5	25	6686	7106.67	−420.67	−1.15
6	25	7270	7106.67	163.33	0.45
7	50	6368	6672.49	−304.48	−0.83
8	50	6691	6672.49	18.52	0.05
9	250	4966	5070.19	−104.19	−0.28
10	250	5189	5070.19	118.81	0.32
11	1000	3085	3154.95	−69.95	−0.19
12	1000	3226	3154.95	71.05	0.19
13	2500	2060	2102.01	− 41.01	−0.11
14	2500	2142	2101.01	40.99	0.11
15	4000	1752	1804.99	− 52.99	−0.14
16	4000	1858	1804.99	53.01	0.14

STD ERROR OF THE ESTIMATE OF Y FROM X = 366.668.
REDUCED CHI SQUARED = 1.1893. Target is 1.0.

Table 1.11 —Thyroxine data.

'x'	1st count	2nd count	Mean count
0	7793	7859	7826
10	7119	6996	7058
25	6232	6344	6288
50	5440	5376	5408
75	4572	4428	4500
100	3846	4081	3964
125	3545	3493	3519
150	3188	3320	3254
200	2708	2774	2741
300	2038	2104	2071

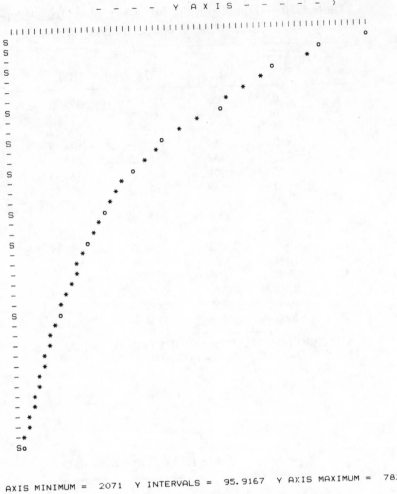

Fig. 1.5 — Computer-generated plot of thyroxine RIA data, processed by the cubic spline program (smoothing factor = 6.31).

series of ten unknowns were reported and we have used the means of these to calculate the unknown concentrations for smoothing factors of 6.31, 0 and 11. The corresponding results for the unknowns from the data processing methods B, C, D and E are shown, as also is the concensus mean for these four methods, in Table 1.12. (Results were not reported for the manual method A.)

Table 1.12 — Results for ten unknowns obtained by using the cubic spline program.

CUBIC SPLINE PROGRAM			PUBLISHED METHODS				
S=6.31	S=0	S=11	B	C	D	E	Concensus
59.1	59.1	58.9	56.3	56.9	56.9	57.5	56.9
71.8	70.5	71.8	71.1	71.6	71.6	71.9	71.5
75.5	73.9	75.6	75.5	76.0	75.9	76.2	75.9
99.1	101.1	99.3	103.1	103.2	103.2	102.8	103.1
107.7	109.8	107.8	112.3	112.3	112.3	111.7	112.2
107.9	110.1	108.0	112.6	112.6	112.6	111.9	112.4
121.2	121.6	121.4	125.3	125.2	125.1	111.9	125.0
133.8	134.3	133.3	136.9	136.5	136.5	135.5	136.4
153.9	156.9	152.0	154.0	153.3	153.3	152.1	153.1
168.0	170.9	165.9	166.8	165.8	165.8	164.5	165.7

When these data were analysed by use of a correlation matrix program, Method E was found to give marginally different results. This was probably due to the effect of the 'outlier' result by Method E for Sample 7. The correlations between the cubic spline fit with a smoothing factor of 6.31 and Methods B, C and D were all identical at 0.997, with there being no apparent advantage in using smoothing factors of zero or 11. Thus, given a reasonable data set, the cubic spline model-free data analysis approach can return results of comparable quality to the logit–log and the four-parameter logistic models.

1.4 THE NON-LINEAR CALIBRATION GRAPH: FITTING BY MEANS OF A PARTIAL SIGMOID CURVE

The smoothed cubic spline can usually fit a reasonable curve to the majority of curve types when these are based on fairly unambiguous data sets. The need to interact with the program and provide a smoothing factor, however, can lead to some problems particularly when the data points are unevenly spaced and the data are inherently noisy. In addition, the cubic-spline fit does not provide a line that can be expressed as a simple continuous algebraic function. It is probably this disadvantage that analysts most dislike. For this reason, we have developed an alternative approach in which the equation to a sigmoid curve is used to fit all curves in categories 2–5 (see section 1.1). This is achieved by using only that portion of a sigmoid that lies in the positive quadrant, by including suitable offsets and scaling factors in the equation, and by raising the x term to a power to accommodate variations in the position of the point of maximum curvature.

Consider the general equation of a sigmoid curve:

$$y = \text{SPAN} \left[\frac{1}{(1 + \exp(-kx^{IN}))} \right]$$

When $x=0$, $y=\text{SPAN}/2$.

When x tends to infinity, y tends to SPAN.

When IN lies between 0.5 and 1.0 the early slope of the curve is reduced and when IN lies between 0.5 and 0, the early slope is increased.

Since all the calibration curves in categories 2 to 5 can be transformed into a curve convex with respect to the x-axis by simple algebraic transforms of the original data, it was proposed that the transformed data could be fitted to the positive convex portion of the sigmoid curve by using linear least-squares analysis after further manipulation of the data.

The equation of the positive convex portion of the curve is

$$y = \text{SPAN} \left[\frac{1}{1 + \exp(-Kx^{IN})} \right] - 0.5 \, \text{SPAN}$$

Let $0.5 \, \text{SPAN} = \text{Vertical Offset} = \text{VOS}$

$$\frac{\text{SPAN}}{y + \text{VOS}} = 1 + \exp(-Kx^{IN}) \tag{1.1}$$

$$\ln \left(\frac{\text{SPAN}}{y + \text{VOS}} - 1 \right) = -Kx^{IN}$$

Since

$$\frac{\text{SPAN}}{y + \text{VOS}} - 1 = \frac{\text{VOS} - y}{y + \text{VOS}}$$

then

$$-1 \cdot \ln \left(\frac{\text{VOS} - y}{y + \text{VOS}} \right) = \ln \left(\frac{y + \text{VOS}}{\text{VOS} - y} \right) = Kx^{IN}$$

$$\ln\left[\ln\left(\frac{y + \text{VOS}}{\text{VOS} - y}\right)\right] = \ln K + IN \cdot \ln x$$

Hence a plot of $\ln x$ vs. $\ln\left[\ln\left(\dfrac{y + \text{VOS}}{\text{VOS} - y}\right)\right]$

should have a slope equal to IN and an intercept of $\ln K$ when analysed by conventional linear-regression analysis. This proved to be a practical approach provided that the data were convex with respect to the x-axis and were always rescaled before analysis: (x data scaled to 0 to 400 and the y data scaled to 0 to 50). In addition care had to be taken to avoid division by zero (e.g. a zero arising from $\ln 1 = 0$), taking natural logarithms of zero ($\ln 0 = $ infinity), or taking natural logarithms of negative numbers. All data were analysed by first setting the value of SPAN arbitrarily to 102 and then to 122; [SPAN(1) and SPAN(2) respectively in the program]. A third estimate of the SPAN was found by examining the sums of the squares of the differences (SSDs), between the back-calculated values of y from the first two fits and the observed values of y. Fig. 1.6 illustrates the

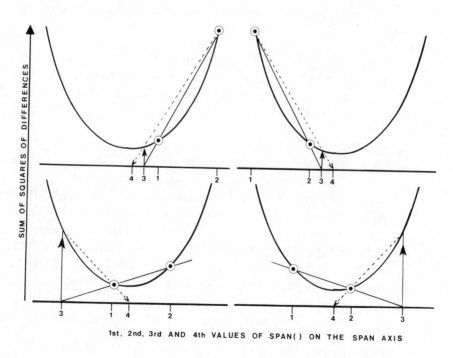

1st, 2nd, 3rd AND 4th VALUES OF SPAN( ) ON THE SPAN AXIS

Fig. 1.6 — Diagrammatic representation of the procedure followed in the partial sigmoid curve-fitting program when searching for the optimum value for the SPAN.

four possible positions of these first two SSD's about the minimum possible, when the SSD's are plotted against the trial values of the SPAN. By drawing a straight line through SPAN(1),SSD(1) and SPAN(2),SSD(2) and then determining its intercept with the y-axis, a third trial value of SPAN [i.e. SPAN(3)], is obtained. The linear regression analysis is repeated and SSD(3) determined. The fastest and most efficient estimate of the next trial value for the SPAN, [i.e. SPAN(4)] is now best found by drawing a straight line between the point with co-ordinates (highest SSD, its associated SPAN) and the point (lowest SSD, its associated SPAN) and taking the resulting intercept on the SPAN axis as the final estimate, SPAN(4), for the fitting procedure.

The above discussion assumed that the data were convex with respect to the x axis. To convert the data from categories 3, 4 or 5 to this form the algebraic transformations used were:

To convert a curve that was concave with respect to the x-axis the rescaled y-values were each subtracted from 50 and the rescaled x-values were each subtracted from 400.

To convert a curve that was concave with respect to both the x- and y-axes the rescaled y-values were each subtracted from 50.

To convert a curve that was convex with respect to both the x- and y-axes, the rescaled x-data were each subtracted from 400.

1.4.1 The CURVEFIT program
The program is divided into thirteen sections and occupies 320 lines of BASIC code, which have been numbered to follow on sequentially from the cubic spline curve-fitting program.

PART A
This includes the data-entry, lines 1428–1504, and array-dimensioning routines, lines 1428–1446. It should be noted that the x and y data must be entered in ascending values of x, and all y replicates entered within their appropriate group. NP is the number of data pairs to be entered. Arrays X(NP) and Y(NP) contain the user's original x and y data, and the XR(NP) and YR(NP) arrays contain the rescaled x and y data as calculated by Part C. Arrays LX(NP) and LY(NP) contain the values of the natural logarithm, (ln), of the rescaled x and the

$$\ln\left[\ln\left(\frac{y + \text{VOS}}{\text{VOS} - y}\right)\right]$$ term for passing to the linear regression routine, Part H; note that at this point the rescaled x and y data are used,

viz. XR(NP) and YR(NP). YC(NP) contains the back calculated values of *y* for the curve-fit under current consideration; values are calculated at line 1840.

As outlined in the preceeding discussion, two preset values of SPAN are examined first, followed by two derived values of SPAN: hence there are four arrays each of four elements for storing the associated VOS, SSD, K and IN; line 1438. In line 1446 two additional arrays are dimensioned, MX(RNP) and MY(RNP), to store the mean values of *x* and *y* data, where replicate values of *y* are entered for each value of *x*. These mean *y* values are derived only to facilitate the simple plotting routine for display on the monitor, Part D. If replicate *y*s are entered then an opportunity is given for the user to view the corresponding table of mean *y*s, lines 1482–1512, for a short time as determined by lines 1508–1512.

PART B
This short routine determines the maxima and minima of the *x* and *y* values in the X(NP) and Y(NP) arrays.

PART C
This routine uses the maxima and minima to rescale the data such that the rescaled *x* data lie within the interval zero to 400 and the rescaled *y* data lie in the interval zero to 50.

PART D
This is a very simple plotting routine for displaying the data on the monitor so that the user can identify the form of the curve that should pass through the points. The *x*-axis is 21 lines and the *y*-axis is 51 characters. The graph has the intercept of the *x* and *y* axes in the top left hand corner, and the *y*-axis runs from zero to 50 down the left-hand edge of the screen. (Should the user prefer a more elegant graphical representation then the routine given in lines 408 to 468 of the linear-regression program could be adapted and inserted here as an alternative.) It is esential that the user correctly identifies the shape of the curve that would best fit the data at this point of the program (lines 1630–1644). This identification is stored as Q$; Q$ is not and cannot be used for any other purpose within this program.

PART E
This section consists of five subsections, the first of which (lines 1646–1652), directs the program flow to the appropriate data-transformation routine.

Part E1, lines 1656–1670, transforms data which are concave with respect to both the x- and y-axes to the 'concave with respect to the x-axis form' required.

Parts E2 and E3, lines 1672–1686 and lines 1688–1702 respectively, perform the transformation necessary for input data which are convex with respect to the x- and y-axes or are concave with respect to the x-axis.

PART E4, PART H AND PART I

Part E4 in association with Part H form the kernel of the program in which the curve-fitting parameters are actually determined. Part I provides an assessment of the goodness of fit of the current model equation. In line 1728–1744 the values of the expressions:

ln of the rescaled x data

and

$$\ln \left[\ln \left(\frac{y + \text{VOS}}{\text{VOS} - y} \right) \right]$$

are determined for passing to the linear-regression routine. Line 1730 is important because it detects zero values of XR(I) and places a dummy marker value of zero into the LY(I) array. This in turn signals to the linear-regression routine, at line 1884, to ignore and exclude that point. In addition, line 1730 prevents the program from attempting to take the natural logarithmm of zero at line 1732:

$$\text{LX(I)} = \text{LOG(XR(I))}$$

Line 1734 acts as a trap for values of (VOS−YR(I)) which are equal to zero, which would cause the program to fail at line 1736 if division by zero was attempted, or division by a negative number which would cause the program to fail at line 1742. Again a dummy marker value of zero is placed into the LY(I) array as a signal for line 1884. Line 1740 detects those situations where the value of Y determined at line 1736 has a value of one or less, which would cause program failure at line 1742.

Part H, lines 1874–1906, is a straightforward linear-regression subroutine which returns the values of the slope and intercept to Part

E4, lines 1754 and 1756, which converts these into the estimates of K and IN for the particular value of SPAN being assessed at that time.

Part I, lines 1908–1930, is a subroutine called at line 1760 to provide an assessment of the goodness of fit for the recently determined values of K and IN. Because of rounding errors and errors inherent in the exponential and logarithmic algorithms used by computers, the values of SPAN and VOS are recalculated by lines 1912 and 1914 for the value of K calculated at line 1756. The loop, lines 1920–1926, determines the sum-of-squares of the differences using the rescaled data in the YR(NP) array.

After the two initial trial values of SPAN, 102 and 122, have been evaluated, the program branches at line 1770 to the subroutine at line 1936, Part J, then at line 1774, to the subroutine at line 1952, Part K.

PART J AND PART K
Part J, lines 1934–1946, determines the trial value of SPAN(3) by the method previously described and illustrated in Fig. 1.6.

Part K, lines 1950–1974, determines the final trial value of SPAN, SPAN(4), by extrapolation back to the SPAN axis of the line passing through the points (highest SSD, its associated SPAN) and (lowest SSD, its associated SPAN). The highest and lowest SSD's are determined by lines 1960–1966 from the series of three SSD's so far derived.

Both modules exploit the simple algebra associated with the determination of the slope and intercept of a straight line:

$$\text{SLOPE} = \triangle y / \triangle x$$
$$\text{INTERCEPT} = y - \text{Slope} \cdot x$$

PART F
This gives a report on the goodness of fit of the model equation with the smallest SSD. The report includes the values of SPAN, K and IN for the data in RESCALED form. It is followed by the statistics of the fit to the original data, *viz.* the SSD for the fit to the data, line 1812, and an estimate of the SE of the estimate of *y* from *x*, line 1814. Note that the latter is determined by using the formula:

$$\text{SQR}(\text{SSD}/(\text{NP}-3))$$

where 3 is the number of parameters estimated, (*viz*. SPAN, K and IN).

PART G

This routine back-calculates the values of *y* from the equation of the line of best fit, line 1840. This equation returns values of *y* in rescaled form, so lines 1842 and 1846 are included to convert the results from line 1840 back into the original form. Line 1848 determines the difference betwen the observed *y* and the back-calculated value of *y*. At line 1856 this difference is interpreted as being significant or insignificant according to whether or not the difference is greater than ± 1.96 times the SE of the estimate of *y* from *x*. Line 1862 provides the user with the opportunity to delete one point and recalculate a new equation.

PART M

If a new fit is to be calculated from the edited data, a flag, F, is set equal to one (line 1866), and the subroutine Part M (lines 2014–2040) is called to delete the user-selected data pair. Line 1704 then returns the program to line 1704 in Part E4 to recommence the full curve-fitting process . The inclusion of the flag, F, in lines 1862–1866, prevent these lines from being re-executed after the repeat fit has been completed.

Line 1868 provides the user with an opportunity to calculate values of *x* from given values of *y* by going to Part L, lines 1976–2010. Otherwise, the program stops (2010).

PART L

This section calculates values of *x* from user-supplied values of *y*. Line 1990 converts the input *y* to the rescaled units and line 1992 transforms this result if necessary. Eq. 1.1, has been rearranged as follows to facilitate the calculation of these values of *x*:

$$\frac{SPAN}{y + VOS} = 1 + \exp(- Kx^{IN})$$

$$\ln\left(\frac{SPAN}{y + VOS} - 1\right) = - Kx^{IN}$$

Let

$$\ln \left(\frac{\text{SPAN}}{y + \text{VOS}} - 1 \right) = \text{BRACKET} \ \text{.......} \ \text{(see line 1994)}$$

Hence

$$\ln \left(\frac{-\text{BRACKET}}{K} \right) = IN.\ln x$$

$$\ln x = \frac{1}{IN} \ln - \left(\frac{-\text{BRACKET}}{K} \right)$$

$$= \text{LNANS} \ \text{........(see line 1996)}$$

$$x = \text{EXP(LNANS)} \ \text{.....(see line 1998)}$$

Line 2000 transforms the resulting value of x, if necessary, and line 2002 rescales the answer.

1.4.2 Listing of CURVEFIT

```
1400 ' CURVEFIT.BAS
1402 ' CURVEFIT -" GENERALIZED CURVE FITTING PROGRAM
1404 ' T F HARTLEY
1406 ' COMPUTERIZED QUALITY CONTROL
1408 ' PUBLISHED BY ELLIS HORWOOD, ENGLAND, 1986
1410 ' .............................................
1412 '
1414 FOR I = 1 TO 5 : PRINT : NEXT I
1416 '
1418 PRINT "*   *   *   *    GENERALIZED CURVE FITTING PROGRAM *   *   *   *"
: PRINT1420 PRINT " CURVEFIT.BAS"
1422 PRINT : PRINT
1424 '
1426 'PART A .......................................................
1428 'DATA ENTRY ROUTINE
1430 INPUT "NUMBER OF X,Y PAIRS "; NP
1432 INPUT "HOW MANY REPLICATES AT EACH X VALUE ... 1,2,3..."; RP
1434 '
1436 DIM X(NP), Y(NP), XR(NP), YR(NP), LX(NP), LY(NP), YC(NP)
1438 DIM SPAN(4), VOS(4), SSD(4), K(4), IN(4)
1440 '
1442 RNP = NP / RP
1444 '
1446 DIM MX(RNP), MY(RNP)
1448 '
1450 PRINT : PRINT "ENTER YOUR DATA PAIRS NOW" : PRINT
```

```
1452 PRINT "NOTE : ENTER ALL REPLICATES IN STRICT SUCCESSION" :PRINT
1454 '
1456 FOR I = 1 TO NP
1458 PRINT "DATA PAIR NUMBER ";I;"...";
1460 INPUT "X,Y = "; X(I), Y(I)
1462 INPUT " DATA OK ....... RETURN / N "; QQ$
1464 IF QQ$ <> "" THEN PRINT "REENTER DATA ..." : GOTO 1458
1466 PRINT
1468 NEXT I
1470 '
1472 'DETERMINE THE MEANS OF THE REPLICATES
1474 IF RP > 1 THEN 1476 ELSE 1516
1476 K = 1
1478 PRINT
1480 '
1482 FOR J = 1 TO RNP
1484 '
1486 FOR I = K TO (K + RP − 1)
1488 MX(J) = MX(J) + X(I)
1490 MY(J) = MY(J) + Y(I)
1492 NEXT I
1494 '
1496 MX(J) = MX(J) / RP : PRINT "MEAN X =";MX(J); " MEAN Y =";
1498 MY(J) = MY(J) / RP : ?MY(J)
1500 K = K + RP
1502 '
1504 NEXT J
1506 '
1508 'PAUSE FOR 5 SECONDS
1510 PRINT "PAUSING FOR 5 SECONDS ................................."
1512 FOR T = 1 TO 5000 : NEXT T
1514 '
1516 'PART B ....................................................
1518 'DETERMINE THE MAX AND MIN OF THE Y DATA
1520 '
1522 MINY = 1000000 : MAXY = −1000000
1524 MINX = 1000000 : MAXX = − 1000000
1526 '
1528 FOR I = 1 TO NP
1530 IF Y(I) < MINY THEN MINY = Y(I)
1532 IF Y(I) > MAXY THEN MAXY = Y(I)
1534 IF X(I) < MINX THEN MINX = X(I)
1536 IF X(I) > MAXX THEN MAXX = X(I)
1538 NEXT I
1540 '
1542 'PART C ....................................................
1544 'RESCALE X AND Y DATA AND STORE IN XR AND YR ARRAYS
1546 'X DATA SCALED 0 TO 400 Y DATA SCALED 0 TO 50
1548 '
1550 XFACTOR = 400 / ( MAXX − MINX )
1552 YFACTOR = 50/(MAXY − MINY)
1554 '
1556 FOR I = 1 TO NP
1558 XR(I) = XFACTOR * (X(I) − MINX)
1560 YR(I) = YFACTOR * (Y(I) − MINY)
1562 NEXT I
1564 '
```

```
1566 'PART D ....................................................
1568 'PLOT ORIGINAL DATA
1570 '
1572 PRINT " !";
1574 FOR J=1 TO 5 : PRINT ".........!"; : NEXT J
1576 PRINT "Y"
1578 C = NP / RP
1580 '
1582 FOR I = 1 TO C
1584 IF RP = 1 THEN XF = XR(I) : XS = XR(I−1) : Y = YR(I) :
       GOTO 1596
1586 XF = MX(I) : XS = MX(I−1) : Y = MY(I)
1588 XF = XFACTOR * ( XF − MINX )
1590 XS = XFACTOR * ( XS − MINX )
1592 Y = YFACTOR * ( Y − MINY )
1594 IF XS < 0 THEN XTICK = 0
1596 XTICK = (XF − XS) / 20 − 1
1598 IF XTICK <= 0 AND Y > 0 THEN PRINT "−"; : GOTO 1614
1600 IF XTICK <= 0 AND Y = 0 THEN PRINT "−*"; : GOTO 1616
1602 '
1604 FOR J = 1 TO XTICK
1606 PRINT "−"; TAB(53) "−"
1608 NEXT J
1610 '
1612 PRINT "−";
1614 PRINT TAB(Y + 2) "*";
1616 IF RP = 1 THEN PRINT TAB(53) "− "; X(I); Y(I)
1618 IF RP > 1 THEN PRINT TAB(53) "− "; MX(I); MY(I)
1620 NEXT I
1622 '
1624 PRINT "X!";
1626 FOR J = 1 TO 5 : PRINT ".........!"; : NEXT J
1628 PRINT
1630 INPUT "NOTE THE SHAPE OF THIS CURVE THEN PRESS ANY KEY"; QQ$
1632 '
1634 ' PART E ....................................................
1636 ' CALCULATE FIT TO STARTING EQUATION
1638 PRINT "IS CURVE CONCAVE (=1) OR CONVEX (=2) WITH RESPECT TO ";
1640 PRINT "THE X AXIS OR"
1642 PRINT "CONCAVE W. R. T. X AND Y AXES (=3) OR"
1644 PRINT " CONVEX W. R. T. X AND Y AXES (=4) 1 / 2 / 3 / 4 "; :
       INPUT Q$
1646 IF Q$ = "2" THEN GOTO 1706
1648 IF Q$="1" THEN GOTO 1690
1650 IF Q$ = "3" THEN GOTO 1658
1652 IF Q$ = "4" THEN GOTO 1674
1654 '
1656 ' PART E1 ....................................................
1658 'TRANSFORM FROM CONCAVE W. R. T. X AND Y AXES
1660 '
1662 FOR I = 1 TO NP
1664 YR(I) = 50 − YR(I)
1666 NEXT I
1668 GOTO 1706
1670 '
1672 ' PART E2 ....................................................
1674 'TRANSFORM DATA FROM CONVEX W. R. T. X AND Y AXES
```

```
1676 '
1678 FOR I = 1 TO NP
1680 XR(I) = 400 − XR(I)
1682 NEXT I
1684 GOTO 1706
1686 '
1688 ' PART E3 ....................................................
1690 'TRANSFORM DATA FROM CONCAVE TO CONVEX FORM
1692 '
1694 FOR I = 1 TO NP
1696 YR(I) = 50 − YR(I)
1698 XR(I) = 400 − XR(I)
1700 NEXT I
1702 '
1704 ' PART E4 ....................................................
1706 SPAN(1) = 102 : SPAN(2) = 122 : VOS(1) = 51 : VOS(2) = 61
1708 '
1710 'CURVE FITTING ROUTINE STARTS HERE
1712 'CALCULATE K AND IN FOR SPAN = 102, 122, SPAN(3) & SPAN(4)
1714 '
1716 FOR J = 1 TO 4
1718 PRINT STRING$(70, "−")
1720 PRINT "NORMALISED MODEL PARAMETERS FOR FIT #";J
1722 SPAN = SPAN(J)
1724 VOS = VOS(J)
1726 '
1728 FOR I = 1 TO NP
1730 IF XR(I) = 0 THEN LX(I) = 0 : LY(I) = 0 : GOTO 1744
1732 LX(I) = LOG(XR(I))
1734 IF (VOS−YR(I)) <= 0 THEN LY(I)=0 : GOTO 1744
1736 Y = (YR(I) + VOS) / (VOS − YR(I))
1738 IF Y <= 1 THEN LY(I) = 0 : GOTO 1744
1740 '
1742 LY(I) = LOG( LOG(Y))
1744 NEXT I
1746 '
1748 PRINT
1750 GOSUB 1876
1752 '
1754 IN = SLOPE
1756 K = EXP( INTER ) : PRINT "TRIAL K = ";K;" TRIAL IN = "; IN
1758 K(J) = K : IN(J) = IN
1760 GOSUB 1910
1762 '
1764 PRINT "SPAN = "; SPAN; " SSD = ";SSD
1766 SSD(J) = SSD : SPAN(J) = SPAN : VOS(J) = VOS
1768 '
1770 IF J = 2 THEN GOSUB 1936
1772 '
1774 IF J = 3 THEN GOSUB 1952
1776 NEXT J
1778 '
1780 PRINT STRING$(70,".")
1782 INPUT "PRESS ANY KEY TO CONTINUE"; QQ$
1784 '
1786 ' PART F ....................................................
1788 ' DETERMINE LOWEST SSD OF THE 4 CANDIDATE FITS
```

```
1790 '
1792 L = 10000 : LL = 0
1794 '
1796 FOR I = 1 TO 4
1798 IF SSD(I)<L THEN L = SSD(I) : LL = I
1800 NEXT I
1802 '
1804 PRINT STRING$(70, """)
1806 PRINT "NORMALISED MODEL PARAMETERS FOR INFORMATION ONLY :"
1808 PRINT "SPAN = "; SPAN(LL); " K= "; K(LL); " IN= ";IN(LL);
     " SSD = "; SSD(LL)
1810 PRINT : PRINT "STATISTICS OF THE FIT TO YOUR DATA SET :"
1812 CORRSSD = SSD(LL) / (YFACTOR) ↑ 2
1814 SEEST = SQR(CORRSSD / (NP−3))
1816 PRINT "SUM OF SQUARES OF DEVIATIONS FROM THE CALCULATED FIT =
     "; CORRSSD
1818 PRINT "STD ERROR OF THE ESTIMATE OF Y FROM X = "; SEEST
1820 PRINT
1822 '
1824 ' PART G .....................................................
1826 'BACK CALCULATE VALUES OF Y FOR THE GIVEN VALUES OF X
1828 'AND COMPARE WITH THE OBSERVED VALUES OF Y
1830 '
1832 PRINT TAB(6) "";
1834 PRINT "X"; TAB(12) "OBS Y"; TAB(24) "CALC Y"; TAB(39) "DIFF";
     TAB(53) "DIFF/SE OF ESTIMATE"
1836 '
1838 FOR I = 1 TO NP
1840 YC = SPAN(LL)*(1 / (1 + EXP( −K(LL)*XR(I) ↑ IN(LL) ))) − VOS(LL)
1842 IF Q$ = "1" THEN YC(I) = ((50 " YC)/YFACTOR) +MINY
1844 IF Q$ = "2" OR Q$ = "4" THEN YC(I) = (YC / YFACTOR) + MINY
1846 IF Q$ = "3" THEN YC(I) = ((50−YC) / YFACTOR) + MINY
1848 DIFF = Y(I) − YC(I)
1850 PRINT I;
1852 PRINT TAB(5) X(I); TAB(12) Y(I); TAB(24) YC(I); TAB(39) DIFF;
1854 PRINT TAB(58) " "; : PRINT USING "##.##"; DIFF/SEEST;
1856 IF ABS(DIFF/SEEST) > 1.96 THEN PRINT " *" ELSE PRINT
1858 NEXT I
1860 '
1862 IF F<>1 THEN PRINT "DO YOU WANT TO DELETE ONE OUTLIER AND ";:
     INPUT "RECALCULATE ..... Y/N "; QQ$
1864 IF QQ$ = "Y" AND F<>1 THEN GOSUB 2014
1866 IF QQ$ = "Y" THEN F = 1 : GOTO 1704
1868 PRINT "DO YOU WISH TO CALCULATE VALUES OF X FROM GIVEN "; :
     INPUT "VALUES OF Y ..... Y / N "; UK$
1870 IF UK$ = "Y" THEN GOTO 1978 ELSE 2010
1872 '
1874 ' PART H .....................................................
1876 'LINEAR REGRESSION ANALYSIS MODULE
1878 SXY = 0 : SX2 = 0 : N = 0 : SX = 0 : SY = 0
1880 '
1882 FOR I = 1 TO NP
1884 IF LY(I) = 0 THEN GOTO 1894
1886 N = N + 1
1888 SXY = SXY + LX(I) * LY(I)
1890 SX = SX + LX(I) : SY = SY + LY(I)
1892 SX2 = SX2 + (LX(I)) ↑ 2
```

```
1894 NEXT I
1896 '
1898 XM = SX/N : YM = SY/N
1900 SLOPE = (SXY − N * XM * YM) / (SX2 − N * XM * XM)
1902 INTER = YM − SLOPE * XM
1904 RETURN
1906 '
1908 ' PART I .................................................
1910 ' SUM OF SQUARES OF DIFFERENCES = (Y OBSERVED − Y CALC) ↑ 2
1912 SPAN = 50 / (( 1 / ( 1 + EXP(−K*400 ↑ IN))) − 0.5)
1914 VOS = SPAN / 2
1916 SSD = 0
1918 '
1920 FOR I = 1 TO NP
1922 Y = SPAN * ( 1 / (1 + EXP(−K*XR(I) ↑ IN ))) − VOS
1924 SSD = SSD + (YR(I) " Y) ↑ 2
1926 NEXT I
1928 '
1930 RETURN
1932 '
1934 ' PART J .................................................
1936 'DETERMINE SPAN(3) MODULE
1938 M = (SSD(2) − SSD(1)) / (SPAN(2) − SPAN(1))
1940 C = SSD(2) − M * SPAN(2)
1942 SPAN(3) = − C/M
1944 VOS(3) = SPAN(3) / 2
1946 RETURN
1948 '
1950 ' PART K .................................................
1952 'DETERMINE SPAN(4) MODULE
1954 'IDENTIFY HIGHEST AND LOWEST SSD OF THE 3 DETERMINED SO FAR
1956 H = 0 : L = 10000 : HH = 0 : LL = 0
1958 '
1960 FOR I = 1 TO 3
1962 IF SSD(I)>H THEN H = SSD(I) : HH=I
1964 IF SSD(I)<L THEN L = SSD(I) : LL=I
1966 NEXT I
1968 '
1970 IF SPAN(HH)SPAN(LL) THEN M = (H−L) / (SPAN(HH) − SPAN(LL)) :
     C = H − M*SPAN(HH) : SPAN(4) = −C/M : VOS(4) = SPAN(4)/2 :
     RETURN
1972 IF SPAN(LL)SPAN(HH) THEN M = (L−H) / (SPAN(LL) − SPAN(HH)) :
     C = L − M*SPAN(LL) : SPAN(4) = −C/M : VOS(4) = SPAN(4)/2 :
     RETURN
1974 '
1976 ' PART L .................................................
1978 'CALCULATION OF UNKNOWN X's MODULE
1980 '
1982 INPUT "HOW MANY UNKNOWNS TO CALCULATE ";UK
1984 '
1986 FOR I = 1 TO UK
1988 INPUT "OBSERVED Y = ";Y
1990 YR = YFACTOR * ( Y − MINY )
1992 IF Q$ = "1" OR Q$ = "3" THEN YR = 50 − YR
1994 BRACKET = LOG( (SPAN(LL) / (YR + VOS(LL) )) − 1)
1996 LNANS = ( 1 / IN(LL) ) * LOG( −1 * BRACKET / K(LL) )
1998 XR = EXP( LNANS )
```

```
2000 IF Q$ = "1" OR Q$ = "4" THEN XR = 400 - XR
2002 X = X(1) + XR / XFACTOR
2004 PRINT "CALCULATED X = ";X
2006 NEXT I
2008 '
2010 STOP
2012 '
2014 ' PART M ....................................................
2016 ' DELETION OF AN OUTLIER ROUTINE
2018 '
2020 INPUT "INDEX NUMBER OF THE PAIR TO BE DELETED ";IN
2022 NP = NP - 1
2024 '
2026 FOR J = IN TO NP
2028 X(J) = X(J+1) : Y(J) = Y(J+1)
2030 XR(J) = XR(J+1) : YR(J) = YR(J+1)
2032 LY(J) = LY(J+1) : LX(J) = LX(J+1)
2034 YC(J) = YC(J+1)
2036 NEXT J
2038 '
2040 RETURN
```

1.4.3   Curve fitting by using the partial sigmoid equation: worked examples

The same test data that were used to evaluate the cubic spline curve-fitting program were used to evaluate the performance of this program; the sigmoid curve data (Table 1.7), were not used as these were inappropriate. The hyperbolic data shown in Table 1.4 produced a SE of the estimate of y from x of 1.24, which was better than the spline fit for a smoothing factor of 6.31, but not quite as good as with a smoothing factor of zero. Table 1.13(a) presents x-values back-calculated for the same y-values as in Tables 1.5 and 1.6. These results were not as good for the highest value of y, 99.2, and this illustrates the principal drawback with this approach to curve fitting : THE PARTIAL SIGMOID EQUATION CONSTRAINS THE FIT THROUGH THE MAXIMUM AND MINIMUM OF THE Y DATA. (Note that this feature becomes obscured if, during the data point deletion stage, MAXY or MINY are deleted; see Table 1.14.) With the cubic spline fit, a better fit was obtained with smoothing factor of zero, which constrained the fit through the means of the duplicate 'observed' y data. Use of the partial sigmoid equation for the same mean y's, (calculated y in Table 1.4), produced the results shown in Table 1.13(b). These results were a considerable improvement on those shown in Table 1.13(a). Also, the SE of the estimate of y from x was 0.18, which was considerably better than the corresponding value of 1.12 obtained when the cubic spline with a zero smoothing factor was used.

When this program was used to fit the data in Table 1.10 (the IgE assay) the SE of the estimate of y from x was 442, which was worse than the value of 367 obtained from the cubic spline program with the smoothing factor of 4.76. However, the first data point was marked with an asterisk indicating

Table 1.13 — Theoretical and calculated values of x for the hyperbolic curve when fitted by using the partial sigmoid equation program (a) when the data were entered as 'observed' duplicates of y (b) when the data were entered as the means of the duplicates (see Table 1.4).

Observed x	Theoretical x	x calculated from the fit	
	(a)		
45.1	1.500	1.446	(− 3.6%)
75.3	3.496	3.497	(0%)
88.9	5.496	5.507	(0.2%)
95	7.489	7.392	(1.2%)
99.9	12.07	10.60	(−12.2%)
	(b)		
45.1	1.500	1.460	(−2.6%)
75.3	3.496	3.522	(o.7%
88.9	5.496	5.521	(0.5%)
95	7.489	7.516	(0.4%)
99.2	12.07	12.22	(1.2%)

that it was a possible outlier:

$x=0$ Observed $y=7355$
Calculated $y=8397$ Difference $= -1042$
Difference/SE of the estimate $= -2.36$

Subsequent deletion of this first point and reprocessing of the data produced a much improved SE of the estimate of y from x of 348, which was better than the fit in Table 1.10. On reprocessing this edited data set, point 10, 8397 emerged as an outlier with difference/SE of estimate $= 2.46$. The cubic spline program had also identified this point as a possible outlier: difference/SE of estimate $= 2.79$. Table 1.14 documents the results of the fit to the edited data set.

The data from Table 1.11 returned an SE of the estimate of y from x of 120. The cubic spline curve-fitting program, with a smoothing factor of 6.31, gave an SE of the estimate of y from x of 83. The first analysis by the partial sigmoid equation program indicated that the 16th data pair, $(150, 3320)$, was an outlier; difference/SE of estimate $= 1.99$. Deletion of this point and reprocessing of the data made some improvement to the SE of the estimate

Table 1.14 — Data from the IgE assay and results of the partial sigmoid curve-fitting program after the first data point (see Table 1.10) had been deleted.

#	OBS x	OBS y	CALC y	DIFF	DIFF/SE OF ESTIMATE
1	0	7832	8397	−565	−1.62
2	10	7699	7540.53	158.47	0.46
3	10	8397	7540.53	856.47	2.46**
4	25	6686	7102.98	−416.98	−1.20
5	25	7270	7102.98	167.02	0.48
6	50	6368	6638.07	−270.07	−0.78
7	50	6691	6638.07	52.93	0.15
8	250	4966	4969.82	− 3.82	−0.01
9	250	5189	4969.82	219.18	0.63
10	1000	3085	3054.54	30.46	0.09
11	1000	3226	3054.54	171.46	0.49
12	2500	2060	2057.23	2.77	0.01
13	2500	2142	2057.23	84.77	0.24
14	4000	1752	1752	0	0
15	4000	1858	1742	106	0.30

of y from x; it was reduced to 113 but this was still not as good as for the cubic spline fit. The complete results report for this fit to the edited data is shown in Table 1.15. Calculation of the concentrations in the ten unknowns gave

Table 1.15 — Thyroxine RIA data fitted by using the partial sigmoid curve-fitting program after the deletion of the 16th data pair (150, 3320).

#	OBS x	OBS y	CALC y	DIFF	DIFF/SE OF ESTIMATE
1	0	7793	7859	− 66	−0.59
2	0	7859	7859	0	−0
3	10	7119	7058.88	60.12	0.53
4	10	6996	7058.88	− 62.88	−0.56
5	25	6232	6274.22	− 42.21	−0.37
6	25	6344	6274.22	69.79	0.62
7	50	5440	5274.33	165.67	1.47
8	50	5376	5274.33	101.67	0.90
9	75	4572	4507.42	64.58	0.57
10	75	4428	4507.42	− 79.42	−0.70
11	100	3846	3908.50	− 62.50	−0.55
12	100	4081	3907.50	172.50	1.53
13	125	3545	3439.20	105.81	0.94
14	125	3493	3493.20	53.80	0.48
15	150	3188	3071.22	166.78	1.04
16	200	2708	2555.88	152.12	1.35
17	200	2774	2555.88	218.12	1.94
18	300	2038	2038.00	0	0
19	300	2104	2038.00	66.00	0.59

the results shown in Table 1.16. It appears that the partial sigmoid curve-

Table 1.16 — Results for unknowns obtained by using the partial sigmoid curve-fitting program and the data for the standard curve shown in Table 1.15.

	Partial sigmoid curve-fitting program	Cubic spline program with smoothing factor = 6.31	Concensus value
	56.3	59.1	56.9
	70.1	71.8	71.5
	74.1	75.5	75.9
	98.4	991	103.1
	106.4	107.7	112.2
	106.6	107.9	112.4
	117.4	121.2	125.0
	127.1	133.8	136.4
	141.5	153.9	153.1
	152.2	168.0	165.7
MEAN	105.0	109.8	11.2
SD	31.2	35.5	35.6
SEM	9.9	11.2	11.3

fitting returned consistently lower results for the last three unknowns, and the correlation coefficients and linear-regression equations based on the data in Table 1.16 supported this observation:

$y=0.879$ Concensus Value + 7.29 $r = 0.9999$
$y=0.876$ Cubic Spline Fitted Value + 8.78 $r = 0.9956$

where y = the result of back-calculating the concentration, x, using the partial sigmoid equation fit.

Compare this with the linear regression analysis of the cubic spline derived data versus the concensus values:
Cubic Spline Fitted Value of $y=0.995$
Concensus Value = 0.90 $r = 0.9970$

Reprocessing the data with only the mean counts from Table 1.11 did not improve the fit, and it was concluded that the program could not improve its performance in the regions of the 150 and 200 standards. This was probably because of the relatively low gradient of the curve at those points.

2

Batch quality control

In this chapter we are concerned with determining whether or not the results obtained for the quality-control specimens within a batch of samples are truly representative of the 'quality' of the whole batch. This is a particularly difficult decision to make if, for example, the analytical procedure is used infrequently, e.g. once a month or less, so that there is not a great deal of quality control 'history' known about how the method performs. Under these cicumstances the acceptance or rejection of a batch has to be decided almost solely by reference to within-batch performance criteria.

2.1 IMPACT OF PROBABILITY THEORY ON BATCH ACCEPTANCE AND REJECTION

In the industrial production situation, small items such as springs are produced as a single large batch of several thousand from a single coil of wire. The statistics associated with determining the probable quality of such a large batch has been the subject of many years of study and appropriate strategies are well defined. Two of the interesting points to arise from the statistics of 'small sample inspection of the quality of large batches' are:

(i) The small samples that are inspected are large, typically 80 to 200, compared the number of quality controls, QC's, most analytical chemists are accustomed or prepared to include in a batch.

(ii) Confidence of quality is directly related to the size of the sample taken for inspection and not to the batch size. It follows from this that quality control of a large production run, i.e. a large batch, is more economical than that of a small production run, because for each the same number of samples must be inspected to be assured of a specified level of quality.

In a typical analytical laboratory, batch quality control is operated along the following lines:

Batch Size:	15 to 30 samples analysed one-off
Quality control samples:	2 to 4 analysed as one-off
Standards:	3 to 6 analysed in duplicate
Acceptance criterion:	Batch is acceptable if the results for the QC's are within ±2 SD of the value stated by the manufacturer of the QC material

If the probabilities of various typical batch quality control schemes and their outcomes are examined, some interesting features become apparent.

Suppose that the PROBABILITY of a QC result being within the range (stated value ±2 SD) is 0.95.

SCHEME A
2 QC's per batch
Acceptance Rule: Both QC's must be within range; (range = the mean ±2 SD).
Probability of Acceptance = $0.95 \times 0.95 = 0.903$
Consequence: Almost 10% of batches will be rejected for failing this rule; (probability of rejection = $1 - 0.903 = 0.097$)

SCHEME B
2 QC's per batch
Acceptance Rule: Both QC's or any one must be within range.
Probability of Acceptance = $\quad 0.95 \times 0.95$
$$+0.95 \times 0.05$$
$$+0.05 \times 0.95$$
$$=0.998$$
Consequence: Less than 1% of batches will be rejected for failing this rule; (probability of rejection = $1 - 0.998 = 0.002$)

SCHEME C
3 QC's per batch
Acceptance Rule: All three QC's within range.
Probability of Acceptance $= 0.95 \times 0.95 \times 0.95 = 0.857$

Consequence: Aproximately 14% of batches will be rejected for failing this rule; (probability of rejection $= 1 - 0.857 = 0.143$)

SCHEME D
3 QC's per batch
Acceptance Rule: All three QC's or any two must be within range.
Probability of Acceptance $=0.95 \times 0.95 \times 0.95$
$$+0.95 \times 0.95 \times 0.05$$
$$+0.95 \times 0.05 \times 0.95$$
$$+0.05 \times 0.95 \times 0.95$$
$$=0.992$$
Consequence: Less than 1% of batches will be rejected for failing this rule; (probability of rejection $= 1 - 0.992 = 0.008$)

SCHEME E
4 QC's per batch
Acceptance Rule: All four QC's must be within range
Probability of Acceptance $=0.95 \times 0.95 \times 0.95 \times 0.95$
$$=0.815$$
Consequence: More than 18% of batches will be rejected for failing this rule; (probability of rejection $= 1 - 0.815 = 0.185$)

SCHEME F
4 QC's per batch
Acceptance Rule: All four QC's or any three must be within range.
Probabilty of Acceptance $=(0.95)^4$
$$+4 \times (0.95 \times 0.95 \times 0.95 \times 0.05)$$
$$=0.986$$
Consequence: Just over 1% of batches will be rejected for failing this rule; (probability of rejection $= 1 - 0.986 = 0.014$)

From the above examples it is apparent that
(i) The more QC's used, the greater the likelihood of rejecting a batch.
(ii) The 'high' rejection probability is substantially reduced by relaxing the acceptance rule to permit one or more of the QC's to be out of range. The formulae below permit the reader to calculate the probability of acceptance for any number, n, of QC's with any number, m, of these being permitted to fall outside the stated (mean ± 2 SD) limits:

Probability of all n QC's within range $= (0.95)^n = P_{all}$

Probability of m QC's within range and $n-m$ QC's out of range
$$= (0.95)^m \, (0.05)^{n-m} \, C_{n,m}$$
$$= P_{\text{partial}}$$

Probability of Acceptance $= P_{\text{all}} + P_{\text{partial}}$

Note: $C_{n,m}$ = the number of combinations of m within-range
QC results from a scheme involving n QC samples

$$= \frac{n\,!}{m\,!\,(n\,-\,m)\,!}$$

So far we have considered only half the problem: we have described the combination of probabilities when the analytical method is performing correctly and therefore in each batch 95% of results for each of the QC's do fall within the appropriate (mean ± 2 SD) acceptance range. We now need to determine what inference/extrapolation about the quality of the whole batch can be made from such an observation. For example if we have found one, two, three, ... QC's to be of an acceptable quality what are the odds that the rest of the batch are of equivalent quality.

(Analogy: An engineer purchases a box of 100 ball bearings which are stated to be 10 ± 0.06 mm in diameter. He takes out three at random, measures them and finds that they are 9.94, 9.98 and 10.05 mm in diameter. What are the chances that the remaining 97 ball bearings are also within the ± 0.06 mm limit?)

This aspect of the problem is addressed via the Operating Characteristic Table as shown in Table 2.1, which was produced by the program included in Appendix C. The figures in Table 2.1 give the percentage chance of accepting a batch with a prescibed proportion of defectives in the batch (see [10, 11]). If P is the proportion defective, (line 9242, program in Appendix C), then the proportion that are not defective is Q = (1−P), (line 9264 program in Appendix C) and the probability of acceptance is PA=Q^N,(line 9264, program in Appendix C) where N is the number of items in the QC sampling scheme, ALL of which are within the acceptance limits. Hence the percentage chance of accepting a batch is PA×100%. The data in Table 2.1 reveal that if we have 2 out of 2 QC's in range then there is an 81% chance of ACCEPTING a batch of analyses which COULD have 10% defectives. Conversely there is a 19% chance of rejecting a batch of analyses which could have 10% defectives.

Table 2.1 — Operating characteristic table for batch quality control design

Acceptance criterion is that all the QC's in the batch must fall within the prescribed limits.

No outliers are permitted.

Figures in this Table equal the percentage chance of accepting the batch.

No. of QC's	PROPORTION DEFECTIVE IN THE BATCH, %								
	0.1	1	2	5	10	20	30	40	50
1	99.9	99	98	95	90	80	70	60	50
2	99.8	98	96	90.3	81	64	49	36	25
3	99.7	97	94.1	85.7	72.9	51.2	34.3	21.6	12.5
4	99.6	96.1	92.2	81.5	65.6	41	24	13	6.3
5	99.5	95.1	90.4	77.4	59	32.8	16.8	7.8	3.1
10	99	90.4	81.7	59.9	34.9	10.7	2.8	0.6	0.1
15	98.5	86	73.9	46.3	20.6	3.5	0.5	0	0
20	98	81.8	66.8	35.8	12.2	1.2	0.1	0	0
30	97	74	54.5	21.5	4.2	0.1	0	0	0
40	96.1	66.9	44.6	12.9	1.5	0	0	0	0
50	95.1	60.5	36.4	7.7	0.5	0	0	0	0
100	90.5	36.6	13.3	0.6	0	0	0	0	0

It would appear that we are required to take 30 QC's and find them all within the mean ±2 SD limits if we wish to have a 4.2% chance of accepting a batch which may contain up to 10% defectives; ie. to obtain greater than 95% confidence that we are sending out batches of results with only 10% of them being likely to be incorrect, then we require 30 QC's in the batch!

Clearly we have arrived at a point in the discussion where the extension of the sampling inspection scheme used in the industrial production quality control environment appears to have assumed an air of impracticality, as far as its application to small to medium batches of samples for chemical analysis are concerned. Nevertheless, the message that the our degree of confidence about the quality of a batch is very heavily dependent on the number of observational checks that are made during the analytical procedure, does, and will always, apply.

2.2 THE CUSTOMER'S SPECIFICATIONS FOR BATCH QUALITY AND ITS IMPACT UPON THE ANALYST'S SELECTION OF QUALITY CONTROL LIMITS

What requires more appropriate clarification is the definition of 'proportion defective in the batch' as it applies to analytical work. A true estimate of this quantity can only be made:

    (i) By making a result-by-result comparison between the laboratory results and those obtained for the same samples by an 'absolute' analytical method, e.g. isotope-dilution mass spectrometry, IDMS.

   (ii) By making a result-by-result comparison between the interpretation of the result obtained from your method and the interpretation of the correct result from the 'absolute' method when each is tested against the same low-normal-high scale.

A typical situation could appear as follows:

Interpretation Criteria: LOW<400
                               NORMAL 400−600
                               HIGH>600

Sample A: Concentration measured by laboratory =500, Normal
Sample A: Concentration measured by IDMS   =515, Normal
Sample B: Concentration measured by laboratory =600, Normal
Sample B: Concentration measured by IDMS   =612, High

If the SD on the laboratory measurements was 3 units then it would seem likely that for both samples there was very probably a significant underestimate by the laboratory but this only resulted in a different interpretation of the result when it occurred for Sample B. This leads us to the question : At what point does our customer for these results take action? For example a result of 612, although it is abnormally high, may not warrant any action but but if it had been 615 then some definite action might have been taken by the customer. In summary we have to take three parameters into consideration:

    The Analyst's Allowable Error = AAE
    The Customer's Acceptable Range = CAR
    The Customer's Action Limits = CAL

In batch analysis the aim should be to ensure a probability profile such as that shown in Fig. 2.1, where the probability of a correct result is excellent in the zone between the extremes of the CAR and the CAL values. If, for example the standard deviation of the method was

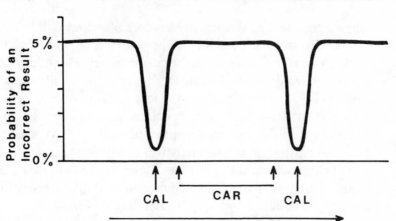

Fig. 2.1 — Ideal error profile of a batch quality control scheme.

$$D = \tfrac{1}{4} \text{ (left-hand CAR} - \text{left-hand CAL)}$$

then we could have a situation like that shown in Fig. 2.2, where a

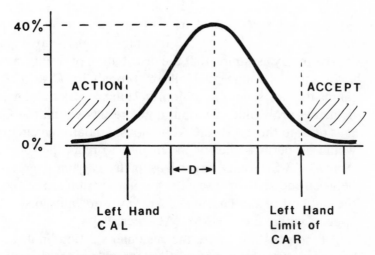

Fig. 2.2 — Probability distribution of results for a specimen with a concentration at the mid-point between the left-hand limit of the CAR and the left-hand CAL when analysed by a method with an SD equal to D.

specimen has a true result that falls exactly at the mid-point between the left-hand limits of the CAL and the CAR. The analytical result would have a 95% chance of falling between these two limits and then being classified as abnormally low, a 2.5% chance of being classified as normal and a 2.5% chance of being low enough for the customer to take action. In this situation the customer in the long run could regard your laboratory as providing 2.5% defective results; he would take action 2.5% of the time when in fact it was not necessary.

The situation, however, becomes more embarassing as the concentration in the specimen approaches the CAL. At that point 50% of the distribution of our analytical results is obliged to fall between the action limit and the left hand limit of the customer's acceptable range. In other words 50% of our results would be falsely higher than the CAL. There is no way of avoiding this because our analytical results at that CAL, or any other point, will always have a Gaussian distribution. Therefore we have to ask the customer what he considers to be an acceptable proportion of false high and low results. If we then assume that the probability distribution of the results is uniform right across the CAL to CAR interval then we can calculate our target standard deviation for the analytical method:

Analyst's 3 SD range =(CAR−CAL) ×Customer's acceptable proportion of false highs or lows

Hence AAE $=\frac{2}{3}$ (CAR−CAL) ×Customer's acceptable proportion of false highs or lows

(Note: By assuming that the distribution of results in the CAL to CAR interval follows a Uniform Probability Distribution we are taking the 'worst case' situation. In clinical work there may well be a definable probability distribution of results within the CAR which tails off into the CAR to CAL regions. Under such circumstances it would be permissable for the analyst to relax the formula for the Analyst's 3 SD range according to the varying degree of overlap between the analytical results' normal distribution at the CAL and the tail end of the customer's known probability distribution as it appears within the CAR to CAL interval.)

In the situation where the customer's action limits are identical with the limits of the customer's acceptable range then of course the above formulae become:

Analyst's 3 SD range =CAR×Customer's acceptable proportion of false highs or lows.

Hence AAE $=\frac{2}{3}$ CAR×Customer's acceptable proportion of false highs or lows

Example 1
CAR=100–200 mmol/l.
CAL=90 and 230 mmol/l.
Customer's acceptable proportion of false lows = 0.10
Customer's acceptable proportion of false highs = 0.05
Analyst's 3 SD range at the left-hand CAL = $(100-90) \times 0.10 = 1.00$
Therefore AAE at the left-hand CAL = ± 0.667 mmol/l.
Analyst's 3 SD range at the right-hand CAL = $(230-200) \times 0.05$
$$= 1.50$$
Therefore AAE at the right-hand CAL = ± 1.00 mmol/l.

Example 2
CAR=100–200 mmol/l.
CAL=100 and 200 mmol/l.
Customer's acceptable proportion of false lows = 0.10
Customer's acceptable proportion of false highs = 0.05
Analyst's 3 SD range at the left-hand CAR limit = $(200-100) \times 0.10$
$$= 10$$

Therefore AAE at the left-hand CAR limit = ± 6.67 mmol/l.
Analyst's 3 SD range at the right-hand CAR limit = $(200-100)$
$$\times 0.05$$
$$= 5$$

Therefore AAE at the right-hand CAR limit = ± 3.33 mmol/l.

Example 3
There is no CAR but the customer will take action if the change between two successive samples from the continuous-flow reactor are different by ± 10 mmol/l. The acceptable proportion of false significant differences is 0.05.

Analyst's 3 SD range = 10×0.05 on the DIFFERENCE
Therefore variance of the difference = $[(10 \times 0.05)/3]^2$
$$= 0.0278$$
Variance of the difference = $2 \times$ variance on each analytical result
Therefore variance of an analytical result = $0.0278/2$
$$= 0.0139$$
Therefore SD on an analytical result = 0.118
Therefore AAE = ± 0.236 mmol/l

2.2.1   Batch quality control: the strategy so far
1. Determine what the customer's requirements are and then calculate the Analyst's Allowable Error.

2. Purchase quality-control materials
   (a) With values that have been assigned to them by a reputable or 'absolute' method [12, 13].
   (b) With values as close as possible to the customer's action limits.
3. Ensure that your method can produce results with the required standard deviation i.e. AAE/2 or better.

2.2.2 Improving the 'count' of quality control 'tests' per batch

We now return to the problem raised by Table 2.1, *viz.* what practical steps can be taken towards increasing the number of quality control 'tests' per batch so as to improve our confidence in overall batch quality? So far we have probably taken too limited a view of what comprises a quality-control test by implying that it only involves the 'quality-control specimen(s)'. From day-to-day practice, analysts are aware that they do not judge the performance of a batch of analyses solely upon the outcome of the quality-control specimens. They are very much aware that the values obtained for the standards, the shape of the calibration curve, the reproducibility of the readings and the general stability of the instrumentation all contribute to their total evaluation of the quality of the batches. Clearly these aspects should be incorporated in some way into the batch quality-control scheme with each such quantitative/objective test being scored as an item in the 'quality assessment' sample taken from the batch. With this approach we can accumulate the 30 or so tests or inspections which Table 2.1 indicates we need to take. If we use the typical batch description given as an example at the beginning of this chapter, the following emerges as typical of small analytical batch quality assessment/control.

Standards
Number: 3 to 6
Acceptance Criterion: None of the readings appear as outliers on the calibration curve
Therefore total QC TESTS COUNT = 3 to 6

Quality control specimens
Number: 2 to 4
Acceptance Criterion: All concentrations are within the AAE of their assigned value(s)
Therefore total QC TESTS COUNT = 2 to 4

Samples
Number: 15 to 30
Acceptance Criterion: Difference between duplicate instrument readings on the same sample $\leqslant 2.16$ standard deviations of the instrument readings.
Therefore total QC TESTS COUNT = 15 to 30

Grand total QC tests count = 20 to 40
By referring back to Table 2.1, rows 8 and 9, we find that we are approaching the lower echelons of statistical respectability when compared with industrial batch production/acceptance quality-control practices, which have been shown to require 30 or more quality control items before a reasonable operating characteristic is reached. The acceptance criterion : 'Difference between duplicate instrument readings on the same sample $\leqslant 2.16$ standard deviations of the instrument readings' was arrived at as follows:

Probability of a result occurring 1.08 SD's from the mean

$$=0.2225$$

Probability of two results at $+1.08$ and -1.08 SD's from their mean

$$=0.2225 \times 0.2225$$
$$=0.05$$

Hence there is a 95% probability that the difference between two instrument readings on the same sample will be within an interval $\leqslant 2 \times 1.08 = 2.16$ SD's; where SD = the standard deviation of the instrument reading e.g. absorbance, cpm, mV etc.

Example
Number of Standards = 8 (4 measured in duplicate)
Number of QC Specimens = 2
Number of Samples = 20
Problem: The customer advises the analyst that one of the results in the last batch was definitely incorrect. What are the chances of this happening for various quality assessment schemes?
Answer: Make the assumption that the customer is correct and therefore set the Proportion Defective per Batch = 1 in 20 = 5% and then refer to Table 2.1.

Scheme	Total QC Count	Probability of Accepting the Batch	Probability of Rejecting the Batch
G: Only Standards and QC's inspected	10	60%	40%
H: Standards, QC's and duplication of first 10 samples inspected	20	36%	64%
J: Standards, QC's and duplication of all 20 samples inspected	30	22%	78%

To complete the analysis of the options we should now take the customer's point of view of each. We continue to presume that 5% is the true proportion defective and therefore if we do five of these batches of 20 samples per annum then it is likely that the false results rates as perceived by the customer would be as follows:

Scheme G: 5 batches$\times$60%$\times$20 samples$\times$5% defective
            =3 false results per 100 samples per annum
Scheme H: 5$\times$36%$\times$20$\times$5%
            ='1.8' false results per 100 samples per annum
Scheme J: 5$\times$22%$\times$20$\times$5%
            ='1.1' false results per 100 samples per annum

2.2.3   What is the effect of permitting outliers into the scheme?
The 'all correct' criterion may well be regarded by some as being too severe, particularly when a particular assay is tedious or requires very expensive reagents. Unfortunately statistics has no respect for these considerations but it can advise us of the likely consequences of taking short cuts in which we accept a certain number of failures amongst our QC tests, e.g. one outlier on the calibration graph. The answer to the question posed at the beginning of this section lies in the sum of the apppropriate number of terms of the binomial distribution; the number of terms summed equals the number of outliers plus one that are going to be permitted into the scheme.

Example
    We have 20 QC tests per batch of analyses and we would like to know the effect of permitting up to 2 outliers into the scheme.

PA=chance of 0, 1 or 2 outliers

$$= Q^{20}+20\ Q^{19}P+190\ Q^{18}P^2$$

=the sum of the first three terms of the binomial distribution.

By assigning successive values to P, the proportion of defectives in the batch, and solving the equation each time, we obtain the data necessary to plot an operating characteristic curve for this scheme, i.e. P versus PA.

To simplify this process we have written a program, OCCURVE, to produce the operating characteristic curve. This program can be used to determine quickly the likely effect of agreeing to a request from the analytical laboratory staff that a particular batch of analyses be accepted and reported even though they have found one or more outliers than are permitted amongst the QC tests.

2.2.4   The operating characteristic curve program
This program occupies 110 lines of BASIC code, numbered 2200–2420, so that it follows on as a unit from the partial sigmoid curve-fitting program.

PART A
This collects the data from the user with prompts as follows

'How many standards are used in the assay?'; line 2244, input N.
The correct response to this should be the number of standards or related items that are inspected in the calibration procedure. Hence although there may be only 4 standards in an assay, these may be taken through the analytical procedure as duplicate preparations, as in the example in the previous section, or perhaps each standard solution is read twice in the spectrophotometer. In both cases the standards count would be 8, provided, of course, that each has been inspected according to a defined acceptance criterion.
'How many QC's are used in a batch?'; line 2246, input equals QC. This should be interpreted in its broadest sense as has been discussed in section 2.2.2. The most obvious and economical way of improving the QC count is to make duplicate measurements on each of the sample solutions and ensure that the differences between these duplicate instrument readings are ≤ 2.16 SD's of of the accepted SD of the instrument measurements. In this way the total QC tests count would equal the number of actual QC samples plus the number of

specimens on which duplicate instrument readings were made.

'Maximum acceptable number of outliers amongst the standards plus QC''s?'; lines 2248–2250, input equals OL. This is simply the total number of failures the analyst is prepared to admit into the whole repertoire of QC tests that the batch is subjected to.

PART B

This section computes the binomial coefficients relevant to the binomial distribution that describes the analyst's QC testing scheme. This concept has already been encountered under Scheme F in Section 2.1, where the term $C_{n,m}$ was defined; this is actually a binomial coefficient. For a given expansion of the equation $(Q + P^N)$ these binomial coefficients are more easily computed from Pascal's Triangle than from solving the factorial-containing expression given for $C_{n,m}$. Pascal's Triangle, shown below, has a structure such that the members of each successive line equal the sums of the two members in the previous line that are diagonally adjacent to them. In addition each line begins and ends with a '1':

N	BINOMIAL COEFFICIENTS
1	1 1
2	1 2 1
3	1 3 3 1
4	1 4 6 4 1
5	1 5 10 10 5 1
6	1 6 15 20 15 6 1
7	1 7 21 35 35 21 7 1

The program computes the complete Pascal's Triangle down to the line equal to the total number of QC tests in the user's scheme; N = NS + QC, line 2268. (Note: Only the Nth line is stored; all previous lines of the Triangle are overwritten.) This is done simply by placing the contents of say the line for N = 3 into the temporary array TC() but in doing so shifting each coefficient one position to the right, lines 2278–2294, and then summing the columns of the source array NC() with those of the TC() array. The results are placed back into the NC() array, lines 2298–2308, and in the process overwrite the previous contents of that array. The process is outlined below:

Array Column No.	1	2	3	4	5
NC() Contents	1	3	3	1	0
SHIFT RIGHT					
TC() Contents	0	1	3	3	1
Sum into NC()	1	4	6	4	1

PART C

This begins by printing out the title and data of the associated operating characteristic curve; lines 2330–2348. The values of the probability of acceptance, PA, are calculated by two loops : lines 2358–2380 and lines 2388–2418. In the first loop the value of P, the proportion defective in the batch, is assigned a value of 0–0.1 in steps of 0.01, line 2358, and at each step the value of PA is computed in the loop lines 2366–2370 by summing the first OL terms of the binomial distribution:

$$Q = 1 - P \dots\dots\text{line 2360}$$
$$\begin{aligned} PA = \ & NC(1) \times Q^N \\ &+ NC(2) \times Q^{N-1} \times P \\ &+ NC(3) \times Q^{N-2} \times P^2 \\ &+ \dots\dots\dots\dots \\ &+ NC(OL) \times Q^{N-OL} \times P^{OL} \end{aligned}$$

A similar loop, lines 2396–2400, is used to calculate PA within the second loop, lines 2388–2418, in which the value of P is stepped from 0.2 to 0.6 in increments of 0.1. Within each of these two major loops, (lines 2358–2388 and 2388–2418), a simple plotting routine is executed, lines 2378 and 2410, so that a plot of the operating characteristic curve is produced. On a standard 80 column printer the probabilty of acceptance axis, (the y-axis), is presented horizontally as 50 columns with each interval being equal to 0.02. The proportion-defective axis, (the x-axis), is presented as 70 lines with each line representative of 0.01 proportion units. The sixteen values of PA calculated by the program are plotted as asterisks.

2.2.5   Listing of OCCURVE

```
2200 'OCCURVE.BAS
2202 'COMPUTERIZED QUALITY CONTROL
2204 'T F HARTLEY
2206 'PUBLISHED BY ELLIS HORWOOD, ENGLAND, 1986
2208 '
2210 FOR I = 1 TO 10 : PRINT : NEXT I
2212 PRINT STRING$(68, "*")
2214 PRINT "OCCURVE.BAS"
```

```
2216 PRINT "OPERATING CHARACTERISTIC CURVE PROGRAM"
2218 PRINT "THIS PROGRAM CALCULATES AND PLOTS THE OPERATING"
2220 PRINT "CHARACTERISTIC OF THE USER'S PROPOSED BATCH QUALITY";
2222 PRINT " CONTROL SCHEME"
2224 PRINT STRING$(68, "*")
2226 '
2228 PRINT "THIS PROGRAM ASSUMES THAT THE DATA FOR THE CALIBRATION"
2230 PRINT "CURVE FORMS PART OF YOUR BATCH QUALITY CONTROL HENCE"
2232 PRINT "OUTLIERS AMONGST THE STANDARDS AND THE QC SAMPLES"
2234 PRINT "ARE COUNTED INTO THE ACCEPTANCE CRITERION"
2236 PRINT
2238 '
2240 'PART A .................................................
2242 '
2244 INPUT "HOW MANY STANDARDS ARE USED IN THE ASSAY "; NS
2246 INPUT "HOW MANY QC'S ARE USED IN A BATCH "; QC
2248 PRINT "MAXIMUM ACCEPTABLE NUMBER OF OUTLIERS ";
2250 INPUT "AMONGST STANDARDS PLUS QC's "; OL
2252 '
2254 'PART B .................................................
2256 '
2258 'PASCALS TRIANGLE
2260 PRINT : PRINT
2262 PRINT "COMPUTING THE BINOMIAL COEFFICIENTS FROM PASCALS ";
2264 PRINT "TRIANGLE" : PRINT
2266 '
2268 N = NS + QC
2270 DIM NC(N+1), TC(N+2)
2272 '
2274 NC(0) = 1 : NC(1) = 1
2276 '
2278 'SHIFT THE CONTENTS OF THE BINOM. COEFFS. ARRAY, NC(), ONE
2280 'PLACE TO THE RIGHT AND INTO TEMPORARY STORAGE IN ARRAY TC()
2282 '
2284 FOR M = 1 TO N-1
2286 TC(0) = 0
2288 '
2290 FOR I = 0 TO M
2292 TC(I+1) = NC(I)
2294 NEXT I
2296 '
2298 'VERTICAL ADDITION OF THE COLUMNS OF THE NC() AND TC() ARRAYS
2300 '
2302 FOR J = 0 TO M + 1
2304 NC(J) = NC(J) + TC(J)
2306 NEXT J
2308 NEXT M
2310 PRINT "BINOMIAL COEFFS ARE : "
2312 '
2314 FOR J = 0 TO M : PRINT NC(J); : NEXT J
2316 PRINT
2318 '
2320 'PART C .................................................
2322 '
2324 'OPERATING CHARACTERISTIC DATA PLUS CURVE
2326 '
2328 INPUT "WHEN THE PRINTER IS READY PRESS ANY KEY ";Q$
2330 LPRINT : LPRINT : LPRINT : LPRINT
2332 LPRINT "OPERATING CHARACTERISTIC : DATA PLUS CURVE "
2334 LPRINT : LPRINT "NUMBER OF STANDARDS = "; NS
2336 LPRINT "NUMBER OF QC's = "; QC
2338 LPRINT "ACCEPTABLE NUMBER OF OUTLIERS = ";OL
```

```
2340 LPRINT : LPRINT "X AXIS = PROPORTION DEFECTIVE IN THE BATCH, P"
2342 LPRINT
2344 LPRINT TAB(25) "PROBABILITY OF ACCEPTANCE, PA − − −>"
2346 LPRINT " P PA";
2348 LPRINT TAB(16) "!";
2350 FOR I = 1 TO 10 : LPRINT "....!"; : NEXT I
2352 LPRINT
2354 PRINT "CALCULATING THE SUM OF THE FIRST "; OL + 1; " TERMS OF"
2356 PRINT "THE BINOMIAL DISTRIBUTION"
2358 FOR P = 0 TO 0.1 STEP 0.01
2360 Q = 1 − P
2362 PA = 0
2364 '
2366 FOR I = 0 TO OL
2368 PA = NC(I) * Q ↑ (N − I) * P ↑ I + PA
2370 NEXT I
2372 PA = ( INT( PA * 100 + 0.5)) / 100
2374 PRINT P; "      "; PA
2376 LPRINT P; " "; PA;
2378 LPRINT TAB(15) "−"; TAB( 16 + 50 * PA ) "*"; TAB (67) "−"
2380 NEXT P
2382 '
2384 FOR T = 1 TO 9 : LPRINT TAB(15 "−"; TAB(67) "−" : NEXT T
2386 '
2388 FOR P = 0.2 TO 0.6 STEP 0.1
2390 Q = 1 − P
2392 PA = 0
2394 '
2396 FOR I = 0 TO OL
2398 PA = NC(I) * Q↑(N−I) * P↑I + PA
2400 NEXT I
2402 '
2404 PA = ( INT( PA*100 + 0.5 )) / 100
2406 PRINT P; " "; PA
2408 LPRINT P; " "; PA;
2410 LPRINT TAB(15) "−"; TAB(16 + 50 * PA) "*"; TAB(67) "−"
2412 '
2414 FOR T = 1 TO 9 : LPRINT TAB(15) "−"; TAB(67) "−" : NEXT T
2416 '
2418 NEXT P
2420 STOP
```

Figures 2.3 and 2.4 are typical operating characteristic curves produced by the program. In Fig. 2.3 the total QC tests count, NS+QC, was set equal to 10 and the acceptable number of outliers equal to one. In Fig. 2.4 the total QC Test counts was equated to 30 with again only one outlier being acceptable. This program has been used to produce the data in Table 2.2 which summarizes the results when just one outlier is permitted amongst 10 to 100 total QC tests in a batch.

In the earlier discussion we observed that 30 QC tests per batch would place us in the situation where we could be 95% confident of rejecting batches that could contain up to 10% defectives. Reference to Table 2.2, (note the asterisked entries) illustrates that permitting just one outlier into this '30 QC tests per batch' scheme has a dramatic

```
NUMBER OF STANDARDS =  8
NUMBER OF QC's       =  2
ACCEPTABLE NUMBER OF OUTLIERS =  1

X AXIS = PROPORTION DEFECTIVE IN THE BATCH, P

                              PROBABILITY OF ACCEPTANCE, PA --->
     P      PA     !....!....!....!....!....!....!....!....!....!....!
     0      1      -                                              *-
     .01    1      -                                              *-
     .02    .98    -                                             * -
     .03    .97    -                                             * -
     .04    .94    -                                          *   -
     .05    .91    -                                        *     -
     .06    .88    -                                      *       -
     .07    .85    -                                    *         -
     .08    .81    -                                  *           -
     .09    .77    -                               *              -
     .1     .74    -                            *                 -
                   -                                              -
                   -                                              -
                   -                                              -
                   -                                              -
                   -                                              -
                   -                                              -
                   -                                              -
                   -                                              -
     .2     .38    -                   *                          -
                   -                                              -
                   -                                              -
                   -                                              -
                   -                                              -
                   -                                              -
                   -                                              -
     .3     .15    -           *                                  -
                   -                                              -
                   -                                              -
                   -                                              -
                   -                                              -
                   -                                              -
                   -                                              -
     .4     .05    -   *                                          -
                   -                                              -
                   -                                              -
                   -                                              -
                   -                                              -
                   -                                              -
                   -                                              -
     .5     .01    - *                                            -
```

Fig. 2.3 — Operating characteristic curve for a batch QC scheme where the total
number of QC tests equals ten and one outlier is permitted.

effect on our 'rejection' confidence which falls to 82%. In fact to gain
the same position of 95% 'rejection' confidence we have to adopt a
scheme of 'one outlier permitted in a total of 45 QC tests'.

```
NUMBER OF STANDARDS =  12
NUMBER OF QC's      =  18
ACCEPTABLE NUMBER OF OUTLIERS =  1

X AXIS = PROPORTION DEFECTIVE IN THE BATCH, P

                          PROBABILITY OF ACCEPTANCE, PA --->)
  P     PA      !....!....!....!....!....!....!....!....!....!....!
  0    1        -
  .01   .96     -                                              *-
  .02   .88     -                                          *  -
  .03   .77     -                                      *      -
  .04   .66     -                                  *          -
  .05   .55     -                              *              -
  .06   .46     -                          *                  -
  .07   .37     -                      *                      -
  .08   .3      -                  *                          -
  .09   .23     -              *                              -
  .1    .18     -          *                                  -
                -
                -
                -
                -
                -
                -
                -
                -
                -
  .2    .01     - *                                           -
                -
                -
                -
                -
                -
                -
  .3    0       -*                                            -
                -
                -
                -
                -
                -
  .4    0       -*                                            -
                -
                -
                -
                -
                -
                -
  .5    0       -*                                            -
                                                              -
```

Fig. 2.4 — Operating characteristic curve for a batch QC scheme where the total number of QC tests equals thirty and one outlier is permitted.

2.3 FIXED AND RELATIVE BIAS ERRORS IN INTERNAL AND EXTERNAL QUALITY CONTROL SCHEMES, AND METHOD COMPARISON STUDIES

This section provides some guidance on investigating fixed and relative bias errors which may become apparent when measured

Table 2.2 — Operating characteristic curve data for various QC schemes where the total number of QC tests, N, in a batch have been varied between 10 and 100 with only one outlier allowed in each scheme.

For convenience the values of PA, the Probability of Acceptance, have been multiplied by 100 to give percentages.

P = Proportion Defective in the batch.

N=	10	15	20	25	30	40	45
P							
.01	100	99	98	97	96	94	93
.02	98	96	94	91	88	81	77
.03	97	93	88	83	77	66	61
.04	94	88	81	74	66	52	46
.05	91	83	74	64	55	40	33
.06	88	77	66	55	46	30	24
.07	85	72	59	47	37	22	17
.08	81	66	52	39	30	16	12
.09	77	60	45	33	23	11	8
.10	74	55	39	27	18*	8	5*
.20	38	17	7	3	1	0	0
.30	15	4	1	0	0	0	0
.40	5	1	0	0	0	0	0
.50	1	0	0	0	0	0	0

N=	50	60	70	80	90	100
P						
.01	91	88	84	81	77	74
.02	74	66	59	52	46	40
.03	56	46	38	30	24	19
.04	40	30	22	17	12	9
.05	28	19	13	9	6	4
.06	19	12	7	4	3	2
.07	13	7	4	2	1	1
.08	8	4	2	1	0	0
.09	5	2	1	0	0	0
.10	3	1	1	0	0	0

values of the QC material fail to agree with those of the manufacturer of the material. Alternatively the laboratory may be participating in an external quality-control scheme, in which its performance appears to be inadequate, or a new analytical method is being developed and the 'new' method results do not agree very well with those from the 'old' method. The simplest investigation of such situations is to plot a graph with the manufacturer's stated values, the QC survey's mean data or the 'old' method data on the x-axis, the laboratory results on the y-axis, and then examine the slope and intercept determined by linear-regression analysis. (For this to be a valid exercise there must be three or more concentration levels in the QC material under investigation. External quality-control schemes usually distribute more than three concentration levels in their programmes, so they do not offer a problem in this respect. New method evaluation studies should involve the comparison of at least 30 specimens with concentrations spanning the full range of concentrations likely to be encountered during routine use.)

In the discussion of the linear calibration data fitting program, it was pointed out that the x-axis is reserved for the independent parameter. The independent parameter in linear-regression analysis is also assumed to be error free. In none of the three instances cited in this section can the x-axis data be considered as 'error free'. This problem has been the subject of study and debate amongst statisticians for many years [14–19]. We have adopted a regression approach [20], based on the 'geometic mean regression' technique of Ricker and others. This is a particularly simple approach to implement as can be seen from the formulae:

Geometric Mean Regression Slope, GMS

$$=(\text{Slope, } y \text{ on } x)/(\text{Slope, } x \text{ on } y)^{1/2}$$

Geometric Mean Regression Intercept

$$=\text{Mean of the } y \text{ Data} - \text{GMS} \times \text{Mean of the } x \text{ Data}$$

We have written a general purpose program, GMREG, which performs geometric mean regression on as many paired columns of data as the user may require. It occupies 148 lines of BASIC program code numbered to follow on from the operating characteristic curve program. It was originally developed for analysis of a set of amino acid data derived from chromatograms that had been read both manually from the chart recorder paper and by an on-line integrator that was connected directly to the colorimeter's output. In that study it was necessary to process 36 columns by 22 rows of data [21]. There were columns for each of the 18 amino acids: (manual result and integrator result), and there was one row per chromatogram. Clearly this is more than is usually required but, with the increasing use of multichannel analysers, the multicolumn capacity feature of the program has been retained as it may well be of use in evaluation and/ or commissioning studies of such instruments.

PART A

This introduces the user to the data-entry format, which is such that data from all rows in columns 1 and 2 are entered first, followed by all rows in columns 3 and 4, and so on; lines 2566–2590. Missing data pairs can be accommodated by entering "$-,-$"; line 2560; this is detected later by line 2680 and consequently skipped over by the statistics module. The input data are stored in the D$() string array, line 2552, and the seventeen statistical results are stored in array A(17). This is temporary storage only, in that all cells are cleared to zero before the next pair of columns of data are processed; lines 2746–2750.

PART B

This part of the program, lines 2608–2642, prints out a table of the user's data. Line 2600 should be noted because this line initializes a Microline 84 printer to print 17 characters per inch instead of the default setting of 10 characters per inch. If another printer is being used then this line, (line 2600), must be changed. Printing with an 80-column printer is NOT allowed by this program.

PART C

This is the statistics package responsible for determining the y-on-x, the x-on-y and the geometric mean regressions for each of the paired columns in the input table array D$(). It commences by printing out the title and headings for the results table, lines 2656–2666. The results table is organized such that the first line contains the results of the statistical analysis of columns one and two of the input table, the second line the results for columns three and four and so on. Line 2670 sets all the transient storage variables required for the calculations to zero. The variables are:

SX = sum of the data in the x data column
SY = sum of the data in the y data column
XX = sum of the x squared data
YY = sum of the y squared data
N   = the count of the number of x,y pairs currently being analysed
XY = the sum of the products of x and y

The variable names in lines 2678–2698 of the program resemble very closely Part D of the linear calibration data fitting program. Similarly lines 2702–2784 resemble Part F of that program. This section, however, contains the additional equations for the calculations of x on y regression and the geometric mean regression:

BXY = slope of the x on y regression, line 2712
AXY = intercept of the x on y regression, line 2716
GMB = slope of the geometric mean regression, line 2772
GMA = intercept of the geometric mean regression, line 2724

Once the statistical analysis has been completed for the current pair of columns, the results are printed out by lines 2726–2742.

PART D
This provides a footnote to the results table to assist the user to understand the abbreviated headings to the columns in the results table.

Table 2.3 gives some data that illustrate the performance and

Table 2.3 — Data used to illustrate the use of the geometric mean regression program

X1	Y1	Y2	Y3	Y4	X5	Y5
100	97	109	100	112	58	50
110	112	125	115	128	108	160
120	123	138	126	141	147	165
130	127	142	130	145	207	245
140	138	155	141	158	245	350
160	163	183	166	186	304	300
170	172	193	175	196	347	365
180	175	196	178	199	397	425
190	192	215	195.	218	566	540
200	200	224	203	227	771	850

interpretation of the results produced by this program. The first five columns contain simulated data, and columns six and seven, headed X5 and Y5, contain some actual method comparison data. These are a small selection of data based on an actual comparison of the results from a manual colorimetric method for the amino acid, hydroxyproline, in urine, column X5, and the corresponding results for the same

ten samples by a column chromatography method in another laboratory. Column Y1 was derived similarly to the data in Table 1.2; i.e.
the data in Tables A1 and A2 were used to produce a randomized
data set with an underlying relationship approximating to:

$$Y1 = 1.00 \times X1 + 0$$

The data in column Y2 were then obtained by multiplying the
contents of column Y1 by 1.12 to produce a data set with an
underlying linear relationship approximating to:

$$Y2 = 1.12 \times X1 + 0$$

Column Y3 contains data produced simply by adding 3 to each item in
column Y1, so the linear relationship was expected to be
approximately:

$$Y3 = 1.0 \times X1 + 3$$

Finally column Y4 was obtained by adding 3 to each item in column
Y2 to produce data with the approximate relationship:

$$Y4 = 1.12 \times X1 + 3$$

It was expected that regression of X1 and Y1 would illustrate the
results to be expected when two methods produce statistically indis-
tinguishable results. Regression of X1 and Y2 would illustrate the
effect of a relative bias error of 12%. Regression of X1 and Y3 would
illustrate a fixed bias between the two methods of 3 units, and
regression of X1 and Y4, a combination of relative and fixed bias
errors.

2.3.1   Lising of GMREG

```
2500 'GMREG.BAS
2502 'GEOMETRIC MEAN REGRESSION
2504 'T F HARTLEY
2506 'COMPUTERIZED QUALITY CONTROL
2508 'PUBLISHED BY ELLIS HORWOOD, ENGLAND, 1986
2510 '
2512 '
2514 FOR I = 1 TO 5 : PRINT : LPRINT : NEXT I
2516 '
2518 PRINT "GEOMETRIC MEAN REGRESSION OF TABULATED DATA"
2520 PRINT "X THEN Y DATA IN COLUMNS ACROSS"
2522 PRINT
2524 '
2526 'PART A .....................................................
```

```
2528 '
2530 'DATA ENTRY ROUTINES
2532 '
2534 PRINT "EXAMPLE DATA TABLE LAYOUT :" : PRINT
2536 PRINT "GLUCOSE UREA"
2538 PRINT "METHOD A METHOD B METHOD P METHOD Q"
2540 PRINT " Xa Yb Xp Yq"
2542 PRINT " 6.78 7.02 3.76 4.01" : PRINT
2544 '
2546 INPUT "HOW MANY COLUMNS IN YOUR TABLE"; NC
2548 INPUT "HOW MANY ROWS IN YOUR TABLE"; NR
2550 '
2552 DIM D$(NC,NR)
2554 DIM A(17)
2556 '
2558 PRINT : PRINT "INPUT DATA PAIRS IN THE X,Y FORMAT 56.7,87.9"
2560 PRINT : PRINT "INPUT MISSING DATA PAIRS AS ",""
2562 PRINT "PRESS RETURN FOR YES OR N FOR NO" : PRINT
2564 '
2566 FOR C=1 TO NC STEP 2
2568 '
2570 FOR R=1 TO NR
2572 PRINT "ROW "; R;" COLUMNS ";C;" & ";C+1
2574 INPUT " XXXXXX,YYYYY = "; A$,B$
2576 PRINT TAB(21) A$;",";B$
2578 INPUT " OK .... "; Q$
2580 IF Q$ = "N" THEN GOTO 2574
2582 D$(C,R) = A$
2584 D$(C+1,R) = B$
2586 NEXT R
2588 '
2590 NEXT C
2592 '
2594 'PART B ....................................................
2596 '
2598 'DATA PRINTOUT ROUTINE
2600 LPRINT CHR$(28)
2602 'SET UP MICROLINE 84 TO PRINT 17 CHARACTERS PER INCH
2604 WIDTH LPRINT 255
2606 PRINT : PRINT : INPUT "************ PREPARE FOR PRINTING "; Q$
2608 LPRINT "INPUT DATA TABLE" : LPRINT
2610 T=0
2612 '
2614 FOR R = 1 TO NR
2616 '
2618 FOR C = 1 TO NC STEP 2
2620 LPRINT TAB(T) D$(C,R);
2622 LPRINT TAB(T+7) D$(C+1,R);
2624 T = T + 16
2626 PRINT " "; D$(C,R);
2628 PRINT " "; D$(C+1,R);
2630 NEXT C
2632 '
2634 LPRINT : PRINT
2636 T = 0
2638 NEXT R
2640 '
2642 LPRINT : LPRINT
2644 PRINT "******** DONE *********"
2646 '
2648 'PART C ....................................................
2650 '
```

```
2652 'STATS PACKAGE
2654 '
2656 LPRINT "RESULTS OF THE STATISTICAL ANALYSIS" : LPRINT
2658 LPRINT TAB(0) "SX";TAB(13) "SY";TAB(26) "XY";TAB(39) "XX";
2660 LPRINT TAB(52) "YY"; TAB(65) "N"; TAB(78) "MX"; TAB(91) "MY";
2662 LPRINT TAB(104)"SDX";TAB(117)"SDY";TAB(130)"BYX";TAB(143)"BXY";
2664 LPRINT TAB(156) "AYX"; TAB(169) "AXY"; TAB(182) "R";
2666 LPRINT TAB(195) "GMB"; TAB(208) "GMA"
2668 '
2670 SX = 0 : SY=0 : XX=0 : YY=0 : N=0 : XY = 0
2672 '
2674 FOR C=1 TO NC STEP 2
2676 '
2678 FOR R=1 TO NR
2680 IF D$(C,R) = "–" OR D$(C,R) = "" THEN GOTO 2698
2682 X = VAL(D$(C,R))
2684 Y = VAL(D$(C+1,R))
2686 SX = SX + X : A(1) = SX
2688 SY = SY + Y : A(2) = SY
2690 XY = XY + X*Y : A(3) = XY
2692 XX = XX + X*X : A(4) = XX
2694 YY = YY + Y*Y : A(5) = YY
2696 N = N + 1 : A(6) = N
2698 NEXT R
2700 '
2702 MX = SX/N : MY = SY/N : A(7) = MX : A(8) = MY
2704 BC = N/(N"1)
2706 SDX = SQR(BC * ( XX/N – (SX/N) ↑ 2)) : A(9) = SDX
2708 SDY = SQR(BC * ( YY/N – (SY/N) ↑ 2)) : A(10) = SDY
2710 BYX = (XY " N*MX*MY) / (XX – N*MX*MX) : A(11) = BYX
2712 BXY = (XY – N*MX*MY) / (YY – N*MY*MY) : A(12) = BXY
2714 AYX = MY – BYX*MX : A(13) = AYX
2716 AXY = MX – BXY*MY : A(14) = AXY
2718 RC = (XY – N*MX*MY) / SQR(( XX – N*MX*MX) * ( YY – N*MY*MY))
2720 A(15) = RC
2722 GMB = SQR( BYX / BXY ) : A(16) = GMB
2724 GMA = MY – GMB*MX : A(17) = GMA
2726 LPRINT : T = 0
2728 '
2730 FOR L = 1 TO 17
2732 LPRINT TAB(T) A(L);
2734 PRINT L;" ";A(L)
2736 T = T+13
2738 NEXT L
2740 '
2742 LPRINT
2744 SX = 0 : SY = 0 : XX=0 : YY = 0 : N = 0 : XY = 0
2746 FOR I = 1 TO 17
2748 A(I) = 0
2750 NEXT I
2752 '
2754 NEXT C
2756 '
2758 'PART D ......................................................
2760 '
2762 'KEY TO THE COLUMN HEADINGS
2764 '
2766 LPRINT
2768 LPRINT"SX = sum of X data SY = sum of Y data XY = sum of X.Y ";
2770 LPRINT " XX = sum of X.X YY = sum of Y.Y N = number of data pairs ";
2772 LPRINT " MX = mean of X data MY = mean of Y data SDX = sd of X ";
2774 LPRINT "data SDY = sd of Y data"
```

```
2776 LPRINT : LPRINT "BYX = slope y on x BXY = slope x on y ";
2778 LPRINT "AYX = intercept y on x AXY = intercept x on y R = ";
2780 LPRINT "correlation coefficient GMB = geometric mean slope ";
2782 LPRINT "GMA = geometric mean intercept" : LPRINT : LPRINT
2784 LPRINT : LPRINT "Ref Hartley & Huber, Lab. Prac. p119, Oct '84"
2786 LPRINT "Ref Biometry by Sokal & Rohlf, pub Freeman, 2nd Ed."
2788 LPRINT : LPRINT : LPRINT
2790 '
2792 PRINT "YOUR DATA ANALYSIS IS NOW AVAILABLE FROM THE PRINTER "
2794 PRINT : PRINT "END ......"
2796 STOP
```

Table 2.4 — Results of the analysis by the geometric mean regression
program of the data in the first five columns of Table 2.3

Columns Regressed Y on X Regression:	X1 and Y1	X1 and Y2	X1 and Y3	X1 and Y4
Relative bias	+0.5%	+12.6%	+0.5%	+12.6%
Fixed bias	−0.92	−0.95	+2.08	+2.05
X on Y Regression:				
Relative bias	−1.2%	−11.8%	−1.2%	−11.8%
Fixed bias	+1.93	+1.89	−1.03	−0.75
Geometric Mean Regression:				
Relative bias	+0.9%	+13.0%	+0.9%	+13.0%
Fixed bias	−1.44	−1.55	+1.56	+1.45

Table 2.4 presents the results produced by the geometric mean
regression program. Relative bias error has been expressed as a
percentage calculated as:

$$\text{Relative bias error} = 100(\text{Slope} - 1)\%$$

Fixed bias has been defined as being identical to the intercept
calculated by the regression analysis. When columns X1 and Y1, X1
and Y2 and X1 and Y3 were regressed, the relative and fixed biases,
according to the y on x regression, appeared to be approximately half
those estimated from the x on y regression. It was interesting to note
that the geometric mean regression suggested that the relative bias
for X1 *vs.* Y1 and X1 *vs.* Y3 was less than 1% and the fixed bias was
less than 1.6 units of the measurement.

As expected from the artificially imposed design of the data in

Table 2.3, the systematic errors between the data in the X1 and Y2 columns and the X1 and Y4 columns were of the order of 12%; in fact they ranged from 11.8–13.0%.

The statistical parameters relating to the analytical significances of the relative or fixed bias have not been incorporated into the program, but it is possible to calculate the standard errors of the slopes and intercepts etc., and all the intermediate values such as the SDs of the X and Y data are available to the user from the results report if it is considered necessary to make these calculations. We think that in most instances analysts prefer to make a judgement on the acceptability or otherwise of a method comparison or QC survey performance by including other factors such as the speed of analysis, costs of reagents, the customers' requirements for certain levels of accuracy and precision etc. In other words, the geometric mean regression provides objective and quantitative information which are only two of a number of considerations contributing to the decision-making process involved in accepting or rejecting methodological or quality-control survey performance.

Turning now to the hydroxyproline method comparison data in columns X5 and Y5, the regression analyses produced the following results:

Y on X Regression:
>   Relative bias = + 2.1%
>   Fixed bias = + 23.3 μmol/l.

X on Y Regression:
>   Relative bias = + 5.1%
>   Fixed bias = − 12.3 μmol/l.

Geometric Mean Regression:
>   Relative bias = + 3.8%
>   Bias = + 18.2 μmol/l.

Clearly, each laboratory's perception of the other's performance would be quite different if they relied solely upon their y on x and x on y regression results. The geometric mean regression results indicate that both laboratories should concentrate upon determining the source of a fixed bias of the order of 18 μmol/l. in their methods. The relative bias errors are a lesser problem.

The data pair 566, 540 in columns X5 and Y5 provides a useful illustration of the ability of the geometric mean regression technique to find the best fit to the data without weighting in favour of the x or y data. In y on x regression the procedure reduces the sum of squares of the differences between the calculated line and the observed y value in the y direction only. In x on y regression the sum of squares of differences between the observed x values and the calculated line of best fit in the x direction is reduced to a minimum. Geometric mean regression, however, very closely approximates the ideal unbiased situation in which the sum of squares of the perpendicular distances of each of the data points from the line of best fit is minimized. These characteristics have been ilustrated in Fig. 2.3.

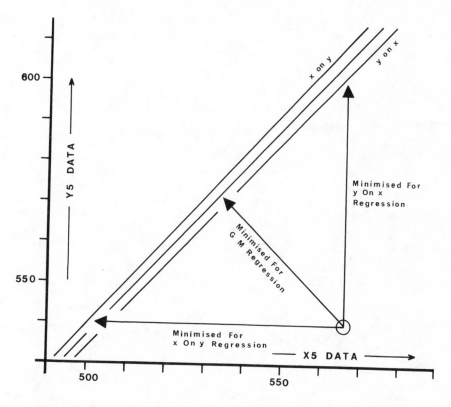

Fig. 2.5 — Illustration of the differences between y on x, x on y and geometric mean regressions in the vicinity of the point 566, 540 from Table 2.3.

2.4 BATCH QUALITY CONTROL: THE FINAL POINTS OF THE STRATEGY

4. Aim to make thirty or more objective quality-control 'tests' per batch of analyses.

5. Determine the operating characteristic of the quality-control scheme particularly when outlier quality control 'tests' are encountered and there is pressure to permit acceptance of the batch.
6. Determine the relative and fixed bias errors in an analytical procedure on a regular basis by using geometric mean regression analysis of the method's data *vs.* data from quality control survey programmes or comparative analyses by other methods or laboratories.

3

Between-batch quality control

3.1 INTRODUCTION

In Chapter 2 the discussion was restricted to the confines of the batch of samples being analysed. In this chapter we will describe techniques, and computer programs, which can be exploited to assess objectively how well our analytical techniques stand the test of time. This involves the closely related activities of trend detection within, and forecasting from, time series of data. Both of these have been subjects of considerable interest to statisticians, economists and engineers (particularly instrumentation and electronic engineers) but all these workers usually have considerably more data in their series than does the average analyst. Consequently the techniques that we have borrowed from their practice are frequently being operated under conditions which might be regarded as suboptimal. Nevertheless, they do provide useful information, and we have generally found them to be valuable aids in between-batch quality-control data reduction. They appear to provide objective signals to augment intuitive feelings about a particular assay's performance at a particular time.

We believe that a balance should be struck between the analyst's professional acumen and computerized quality control. We regard computerization of clerical and computational aspects of quality-control record-keeping as an advantage which permits the analyst to spend more time in the laboratory and less time in the office. However computerization should not become an aid to the remote management of the laboratory from a computer terminal elsewhere. The analyst should be attending to any tasks which the computerized quality-control printout may be highlighting e.g. poor maintenance of volumetric apparatus, deteriorating sensitivity in a spectrophotometer, suspect purity of the current batch of a key reagent, the consequences of an *ad hoc* modification to a procedure etc.

We have computerized three trend-detection devices for between-batch quality control. The first is a more convenient representation of the standard QC chart where the target mean value for the quality-control material plus and minus three standard deviations is presented on the *y*-axis and batch number on the *x*-axis. The computer produces a standard normal deviate plot. For this, the deviations of the QC values from the target mean value are each divided by the target standard deviation, and these transformed data points are plotted on a graph with a *y*-axis scaled in target standard deviation units on either side of a target mean of zero. The trend detection device used with this computerized plot is still the analyst's eye.

The second device is the computerized cusum V-mask [22–24]. Cusums [23,24] are particularly useful in drawing attention to possible shifts in the measured mean for the quality-control material.

The Trigg Tracking Signal, [25] has been adapted for many applications, including the clinical and analytical fields [26]. This technique provides the user with a uniform and objective trend-detection device that is simple to interpret because for all assays the signal is contained within the same window of values of plus and minus one. It does have the disadvantage that if no action is taken to correct the drift away from normality, the tracking signal will move back into the acceptance limits once the trend has stabilized at a new 'set point'. (Some may regard this as ideal behaviour particularly for some of their more temperamental assays – a quality control test that, if it is ignored for long enough, will resolve to accept the new status quo!)

In the following discussion we have chosen to present these tests in the order:

> Cusum V-Mask Scanning
> Trigg Tracking Signal
> Standard Normal Deviate Plot

because this is the order of decreasing complexity in their theory, implementation and programming.

3.2   THE COMPUTERIZED CUSUM V-MASK

The cusum, which is an abbreviation for 'cumulative sum', appears to have been given its first formal description in 1964 [24]. Since then it has been a popular quality-control device in industrial situations and has found applications in the analytical laboratory. The cusum is the cumulative sum of the errors calculated according to the equation:

$$\text{Cusum} = \sum_{i=1}^{n} (x_i - k)$$

where x_i is the ith value obtained on analysis of the quality-control material and k is a reference value often set equal to the target mean. In an assay where there is no sustained drift towards values which are either greater than or less than the reference value, and the values of x_i are normally distributed about the reference value, the cusum result remains close to the reference value because some of the differences (x_i-k) will be positive and some negative. The cusum chart will remain essentially horizontal. However, as soon as drift influences the assay, the cusum assumes a steadily increasing positive or decreasing negative value according to whether the assay is consistently underestimating or overestimating the analyte in the quality control material.

The cusum plot is a chart with a centre value on the y-axis of zero and the batch numbers along the x-axis. An assay that is in control will have a cusum plot with an appearance very similar to the well-known Shewhart control chart. In the Shewhart chart, action is signalled when one or more points fall outside the control limits (e.g. lines at $+3s$ and $-3s$). In a cusum chart, we need to decide when a major change of slope has occurred, and the usual method is to employ a V-shaped mask.

The apex of a clear perspex V-mask overlay is placed a predetermined number, b, of batch numbers ahead of the most recent cusum data point, then it is observed whether any, and if so how many, of the previous cusum points fall outside the limbs of the V-mask. This operation is illustrated in Fig. 3.1; in the top diagram all the previous cusum points lie within the limbs of the V- mask, but in the lower diagram the same data tested with a V- mask with different 'a' and 'b' dimensions indicates that two points fall outside this particular test limit. Because they are below the lower limb of the V-mask then an upward shift in the cusum has been detected by this V-mask and consequently it is providing 'evidence' that the assay has developed a positive bias with respect to the target mean. It can be shown that when the mean shifts by an amount equal to the standard error, an out-of-control signal will be sounded three times more quickly by use of a cusum chart rather than a Shewhart chart [27]. The nature of the

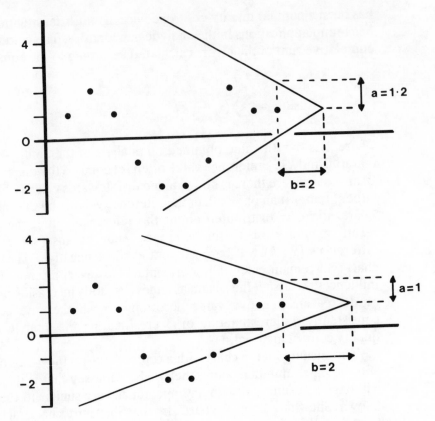

Fig. 3.1 — Two different cusum V-masks used to test identical cusum plots.

confidence/statistical significance ascribed to this evidence depends on the values of *a* and *b* that are employed. We will not want to initiate remedial actions unless there are good grounds for them. The techniques for determining the values of '*a*' and '*b*' for the V-mask based upon conventional statistical considerations have been described [28–30].It should be noted that instead the dimension '*a*', the half angle, θ, between the limbs of the V-mask is often referred to, but:

$$\tan \theta = a/b$$

Thus, it is possible to devise a V-mask with defined statistical properties, but we feel that the methods are somewhat laborious. We do not routinely plot a cusum chart, but we have developed a computerized equivalent. It has been pointed out [23] that the assessment of a time series of data points by means of a linear limits device such as the V-mask could be reduced to a simple pair of test-limit calculations done each time a new data point is added. The formulae are:

$$D_N = \text{Min } (D_{N-1};a) + a/b + e$$
$$R_N = \text{Min } (R_{N-1};a) + a/b - e$$

where

$$e = x_N - x_{N-1}$$

and Min (.........) reads 'take the minimum of either the previous value of D or the V-mask 'a' dimension'. The 'D' test is the test for a downward trend in the data and the 'R' test is the test for a rise (upward trend) in the time series. When either test becomes negative, a point (or points) has fallen outside the upper or lower limb of the V-mask.

Quite how these formulae perform the equivalent of visual inspection is probably not immediately apparent to many readers and we have used the example in Fig. 3.2 as an illustration. A time series

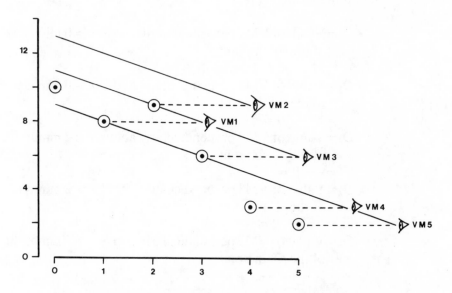

Fig. 3.2 — V-mask used to detect a downward trend in a time series of data points: illustration of the solution developed by Harrison and Davies [23].

of six points is shown and from the '1' th point onwards the upward limb of a V-mask has been drawn with $a = 2$ units and $b = 2$ batches ahead. When we considered the example illustrated in Fig. 3.1 the 'retrospective point of view' that we took of the data was achieved by looking back at the data from the apex of the V-mask. Hence in Fig. 3.2 we have placed 'eyes in profile' at the apex of each V-mask after each data point is added to the series; the apex of each V-mask is indicated by the abbreviations VM1, VM2, etc.. It then becomes apparent that each time we make a new inspection we visually locate and focus in on either that point which is the closest to the upper limb of the V-mask or that point or those points which fall outside the V-mask. We can formalize this visual assessment into two inspection rules:

(i) If no points fall outside the V-mask, identify the point, other than the most recently plotted point, that is closest to the V-mask and then measure and record the distance of that point from the V-mask. Call this distance D.

(ii) If points fall outside the V-mask limb identify which point is the furthest from the limb and measure and record the distance as D.

If we now proceed to apply these rules to the plot in Fig. 3.2, then we obtain the following results:

D_1 = Value of VM1 perpendicularly above $x = 0$ minus the value
    of the '0' th point
= $11 - 10 = 1$
D_2 = Value of VM2 perpendicularly above $x = 0$ minus the value
    of the '0' th point
= $13 - 10 = 3$
D_3 = Value of VM3 perpendicularly above $x = 2$ minus the value
    of the '2' th point
= $9 - 9 = 0$
D_4 = Value of VM3 perpendicularly above $x = 2$ minus the value
    of the '2' th point
= = $7 - 9 = -2$
D_5 = Value of VM5 perpendicularly above $x = 2$ minus the value
    of the '2' th point
= $7 - 9 = -2$

Exactly the same results are obtained using the formulae for D_N and e derived by Harrison and Davies, [23], when the same V-mask parameters of $a = 2$ and $b = 2$ are used:

$$D_1 = 2 + 1 + (8 - 10) = 1$$
$$D_2 = 1 + 1 + (9 - 8) = 3$$
$$D_3 = 2 + 1 + (6 - 9) = 0$$
$$D_4 = 0 + 1 + (3 - 6) = -2$$
$$D_5 = -2 + 1 + (2 - 3) = -2$$

We can therefore be satisfied that these formulae do succeed in reproducing the V-mask and visual inspection procedure, and confidently discard our time series plot and the V-mask.

This procedure was modified by Lewis [22], by replacing 'e' by:

$$e' = x_N - \text{Target Mean}$$

and where

$$e' = e + x_{N-1} - \text{Target Mean}$$

The example in Fig. 3.3 provides a further demonstration that this

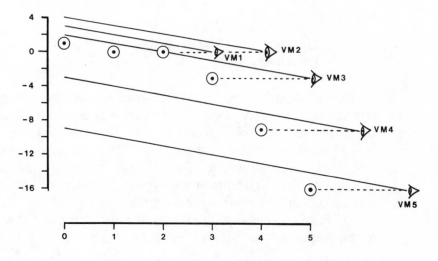

Fig. 3.3 — V-mask used to detect a downward trend in a cusum plot: illustration of the solution used by Lewis [22].

substitution was valid and produces the desired effect. The cusum plot shown in Figure 3.3 is based on the same data as Fig. 3.2 with the Target Mean taken to be 9 in the cusum calculations. By applying the same visual inspection rules, we find:

D_1=Value of VM1 perpendicularly above $x = 0$ minus the value
    of the '0' th point
    $=3 - 1 = 2$
D_2=Value of VM2 perpendicularly above $x = 0$ minus the value
    of the '0' th point
    $=4 - 1 = 3$
D_3=Value of VM3 perpendicularly above $x = 2$ minus the value
    of the '2' th point
    $=0 - 0 = 0$
D_4=Value of VM4 perpendicularly above $x = 2$ minus value of
    the '2' th point
    $=-5 - 0 = -5$
D_5=Value of VM5 perpendicularly above $x = 2$ minus value of
    the '2' th point
    $=-11 - 0 = -11$

Calculations of D with e' substituted for e in the Harrison and
Davies equations, and with $a = b = 2$ as before gave identical results:

$$D_1 = 2 + 1 + (8 - 9) = 2$$
$$D_2 = 2 + 1 + (9 - 9) = 3$$
$$D_3 = 2 + 1 + (6 - 9) = 0$$
$$D_4 = 0 + 1 + (3 - 9) = -5$$
$$D_5 = -5 + 1 + (2 - 9) = -11$$

Hence we can be confident that this new substitution does provide
an equivalent to the V-mask test of a cusum graph, without the need
to actually scale and plot any points onto a cusum chart.

The next procedure requiring simplification/automation is the
selection of the appropriate values of a and b for the V-mask. Lewis
[22] proposed an algorithm that arrived at the V-mask parameters by
simulation, an easy process if a computer is available.

We have adopted the following systematic approach to the search
for an optimum value for the V-mask 'a' dimension:

(i) Establish a disk file containing a run of at least 40 typical
results from the QC material under consideration. These data should
contain one value greater than the mean plus two standard deviations
and one value less than the mean minus two standard deviations of
the assay. Simulated data can be used if necessary.

(ii) Create an artificial upward step, usually +10%, in the series
from the 21st to the 40th data item, inclusive.

(iii) Scan the data with a V-mask that has a fixed value of $b = 2$

and a steadily decreasing 'a' parameter until a step within the interval of the nineteenth to the twentysecond data item is eventually detected. This V-mask is then used prospectively to monitor the QC data obtained for that material.

In our program the V-mask 'a' is set initially at three times the standard deviation of the 40 data items as calculated before the artificial upward step is introduced. As each scan is run the program focuses on the value of R_N which becomes negative whenever the current V-mask detects an increase in the data series. The scan of the series commences by setting the target mean equal to the mean of the first two data points and a significant increase is only substantiated when two successive values, R_N and R_{N-1}, are found to be negative. At that point the process mean is recalculated as the mean of those data items which are associated with the CONTINUOUS series that has the property R_N, R_{N-1}, R_{N-2}, R_{N-3}.... a. In this way the immediate series of data items which have formed the trend towards the upward shift, plus the two items, x_N and x_{N-1} which are definitely implicated, all contribute to the estimate of the new process mean. The latter is substituted for the target mean in the equation for 'e' in the prospective scan from x_{N-1} onwards.

3.3   A PROGRAM TO ESTABLISH A QC DATA FILE AND ANOTHER TO PLACE DATA IN THE FILE

Before we discuss the cusum V-mask scanning program we first outline the arrangement of our QC data files. These are all sequential disk files each with a header of descriptive parameters, such as analyte target mean etc., followed by sufficient array capacity for 100 batch dates each with the associated measured value for the quality control material. They are maintained as circular files in that after the QC results from 100 batches have been entered, the new entries overwrite the first, i.e. the oldest, entries. A file marker is used to keep track of the current situation in the file.

The program QCNAME creates a QC data file with a user-specified name, which can be up to eight characters long, (line 2928) and to which the program adds the file type identifier ".DAT", (line 2930). The file has a header into which the user places the following essential information:

> the analyte, line 2934
> the target mean, line 2936
> the target SD, line 2938

the value for the Trigg Tracking Signal Alpha, line 2940
the value of the cusum V-mask 'a' dimension, line 2942

The user must enter all this information, apart from the V-mask 'a' dimension which is optional at this stage, and is usually written in by the VMASKA program, The Trigg Tracking Signal Alpha value has not yet been described, and the reader is referred forward to Section 3.5 for an explanation and a table of alpha values; we usually use a value of 0.2 for our assays. The FOR NEXT loop (lines 2952–2958), writes dummy dates, 00/00/00, and QC values, zeros, into the D$(100) and VA(100) arrays before writing the whole file consisting of the header (line 2962), and the dummy data (lines 2966–2970), to disk.

QCFILER is a straightforward program which permits the user to write new data into a nominated QC file. Part A (lines 3036–3070), reads the named (N$, line 3028) sequential QC data file and displays the analyte name (line 3048), the target mean (line 3050), the target SD (line 3052), the number of QC entries it contains (line 3054), and the location of the filing marker, FM (line 3056). The filing marker marks the position at which the next date and QC result will be entered into the DA$(100) and VA(100) arrays respectively. For a new file which does not yet contain data from 100 batches the filing marker is always one more than the number of items in the file (line 3108). Once the file contains 100 entries then the filing marker is reset to one (line 3110), and the entry 101 for example overwrites the first entry in the file. Hence to read back the data out of the file in chronological order the read routine must start at the FM th cells of the arrays and read to the 100th entries, then read from the first cells of the arrays to the (FM–1) th cells of the arrays; (for an example, see lines 3892–3908 of the REPORT program).

The first portion of PART B (lines 3072–3114), is concerned with receiving, checking and filing the new entries. The number of entries to be added to the file is obtained from the user (line 3082). The loop 3088–3114 then receives the entries (line 3094), as the batch date ED$ in the format dd/mm/yy and the QC result, R. The entry for R should be restricted to five figures including the decimal point otherwise there could be problems later when the program REPORT is run. The latter is pressed for space when the standard normal deviate plot, the batch date, the Trigg tracking signal and the process mean all have to be accommodated on the monitor within a single 80 column line per batch. The user is given the opportunity to check the date and result entry (line 3098) before proceeding. Lines 3118–3132 then rewrite the entire file back onto the disk.

3.3.1   Listing of QCNAME

```
2900 'QCNAME
2902 ' THIS PROGRAM CREATES A NAMED QC FILE AND FILLS THE DATA
2904 ' ARRAY WITH DATES = 00/00/00 AND DATA = 0
2906 '
2908 'COMPUTERIZED QUALITY CONTROL
2910 'T F HARTLEY
2912 'PUBLISHED BY ELLIS HORWOOD, ENGLAND, 1986
2914 '
2916 DIM DA$(100), E(100)
2918 '
2920 FOR I = 1 TO 5 : PRINT : NEXT I
2922 PRINT STRING$(60, "*")
2924 PRINT " QC FILE NAMING PROGRAM"
2926 PRINT STRING$(60, "*") : PRINT
2928 INPUT "NAME OF NEW QC FILE EG CREAT 8 CHARS ONLY "; N$
2930 N$ = N$ + ".DAT"
2932 OPEN "O", #1, N$
2934 INPUT "ANALYTE = "; A$
2936 INPUT "TARGET MEAN = "; TM
2938 INPUT "TARGET SD = "; SD
2940 INPUT "VALUE FOR TRIGG ALPHA = "; TRA
2942 INPUT "VALUE FOR V-MASK A = "; VMA
2944 NF = 0
2946 FM = 1
2948 PRINT : PRINT "CREATING AND WRITING FILE NOW ...." : PRINT
2950 '
2952 FOR I = 1 TO 100
2954 DA$(I) = "00/00/00"
2956 E(I) = 0
2958 NEXT I
2960 '
2962 WRITE#1, A$, TM, SD, TRA, VMA, NF, FM
2964 '
2966 FOR I = 1 TO 100
2968 WRITE#1, DA$(I), E(I)
2970 NEXT I
2972 '
2974 CLOSE
2976 '
2978 PRINT : PRINT "............ DONE ................." : PRINT
2980 '
2982 STOP
```

Listing of QCFILER

```
3000 'QCFILER
3002 'PROGRAM FOR FILING DATA INTO A QC FILE
3004 '
3006 'COMPUTERIZED QUALITY CONTROL
3008 'T F HARTLEY
3010 'PUBLISHED BY ELLIS HORWOOD, ENGLAND, 1986
3012 '
3014 '
3016 DIM DA$(100), VA(100)
```

```
3018 '
3020 FOR I = 1 TO 5 : PRINT : NEXT I
3022 '
3024 PRINT "............. QC RESULTS ENTRY PROGRAM ................"
3026 PRINT
3028 INPUT "FILE NAME TO USE .... .DAT IS NOT REQUIRED "; N$
3030 N$ = N$ + ".DAT"
3032 PRINT : PRINT
3034 '
3036 ' PART A .....................................................
3038 '
3040 ' READ IN DATA FROM NAMED QC FILE
3042 '
3044 OPEN "I", #1, N$
3046 INPUT#1, ID$, TM, SD, TRA, VMA, NF, FM
3048 PRINT "QC FILE IS "; ID$
3050 PRINT "TARGET MEAN IS "; TM
3052 PRINT "TARGET SD IS "; SD
3054 PRINT "FILE CONTAINS "; NF;" ENTRIES"
3056 PRINT "FILING MARKER IS AT ", FM
3058 '
3060 FOR I = 1 TO 100
3062 INPUT#1, DA$(I), VA(I)
3064 NEXT I
3066 '
3068 CLOSE
3070 '
3072 ' PART B......................................................
3074 '
3076 ' NEW QC RESULTS ENTRY
3078 '
3080 PRINT : PRINT
3082 INPUT "HOW MANY ENTRIES DO YOU HAVE = "; NE
3084 PRINT
3086 '
3088 FOR I = 1 TO NE
3090 PRINT : PRINT "ENTRY # ";I : PRINT
3092 PRINT "DATE ... DD/MM/YY AND RESULT ..."; :
     PRINT " 5 FIGS INCL DEC POINT =";
3094 INPUT ED$, R
3096 PRINT
3098 INPUT "RESULTS OK .............RETURN / N "; Q$
3100 IF Q$ = "N" THEN GOTO 3092
3102 PRINT
3104 DA$(FM) = ED$
3106 VA(FM) = R
3108 FM = FM + 1
3110 IF FM = 101 THEN FM = 1
3112 IF NF = 100 THEN GOTO 3114 ELSE NF = NF + 1
3114 NEXT I
3116 '
3118 OPEN "O", #1, N$
3120 WRITE#1, ID$, TM, SD, TRA, VMA, NF, FM
3122 '
3124 FOR I = 1 TO 100
3126 WRITE#1, DA$(I), VA(I)
3128 NEXT I
3130 '
3132 CLOSE
3134 '
3136 PRINT "FILING COMPLETED ............."": PRINT
3138 STOP
```

We have not written a specific program for file editing and correcting, and usually use this program, QCFILER, to rectify any errors that are missed during data entry. For example if a date and a result are later noted to be incorrect in a file called UREAQC.DAT we run this program as far as line 3082 and at that point we key in CTRL C to halt the program execution. We then use one line BASIC instructions executed in the immediate mode to rectify the error or errors, as in the following sequence, shown as it would appear on the monitor:

```
OK    PRINT DA$(10)
26/1/85
OK    DA$(10) = "26/10/85"
OK    PRINT VA(15)
1.15
OK    VA(15) = 11.5
OK    GOTO 3118
FILING COMPLETED ...........
BREAK IN LINE 3138
```

In the same way information in the header can be changed if necessary.

3.4   THE CUSUM V-MASK SCANNING ROUTINE TO DETERMINE THE OPTIMUM 'a' DISTANCE FOR THE V-MASK

This program, called VMASKA, operates on a QC file that has been loaded previously with forty simulated batch dates and QC results in the first forty positions of the DA$() and VA() arrays. It is envisaged that this program is only run once when a new quality-control material is introduced.

PART A
This reads in the QC file header (line 3256), and the batch dates with their associated QC results into the DA$() and VA() arrays (lines 3276–3280). If the file contains less than the forty entries, line 3270 prints a warning message and halts the program.

PART B
This routine (lines 3288–3320), calculates the standard deviation of the VA$(1 - 40)$ data items, DSD (line 3310), and the mean of these

file data, line 3312. The variables SXX and SX are the sum of the squares of the data items and the sum of the data items respectively.

PART C and PART D

These two sections (line 3322–3478) form the body of the V-mask 'a' dimension search routine.

Lines 3330–3338 effect the artificial upward step in the QC values from the twentyfirst to the fortieth. We usually use a 10% step increment at this point. The program then branches to the subroutine (lines 3590–3664), which forms Part G. The latter is a simple graph-plotting routine identical in style to that used in Part C of the linear calibration graph program. However, instead of printing a graph with 41 rows by 61 columns this routine uses 82 columns. Thus, line 3220 sets the line printer width to 100 columns. If 100 columns cannot be accomodated on the user's printer then the branch at line 3342 will have to be deleted or alternatively the plot squeezed into 42 columns on an 80 column printer.

Lines 3348–3352 define several important variables:

C is a scan counter which is set to an initial value of one.

B is the 'b' dimension of the V-mask which is set to a constant value of two.

U and D are up and down flags which are set equal to one by the tests of D_N and D_{N-1} and R_N and R_{N-1} executed by lines 3392 and 3396 respectively. At line 3400 these flags direct the flow of the program to Part D which makes the appropriate calculations and tests associated with detecting a significant upward or downward change in the process mean.

FLAG takes on values of one or two respectively as a result of tests performed at line 3472 or at line 3692 which is encountered only as a result of branching to Part H of the program from line 3534. For the latter to have happened a 'Y' input must have been made by the user at line 3534.

DECA is the the calculated decrement in the V-mask 'a' that will be used to decrement the value of 'a' set at line 3352 before a new scan is commenced. It is set at 1% of three times the standard deviation of the original data, DSD. This ensures that the value of 'a' reduces only slowly, rather than at the rather rapid rate used by Lewis in his algorithm [22], where the initial value of 'a' was set to 70% of the mean of the first two QC results in the series, VA(1) and VA(2), and thereafter reduced by either 20% or 60% according to how many, or no increments, respectively have been detected in the previous scan of the data by the current V-mask.

Lines 3358 and 3360 inform the user of the scan number that the program is currently executing along with a rapid display of the candidate increases as signalled by values of R() that are less than 'a' but greater than zero, line 3396, and more importantly values of R() that are negative, line 3398. Of course, if two successive values of R() are negative then the position of a significant positive shift in the process mean has been detected by that particular V-mask.

Line 3362 sets the initial value of the target mean, M, to the mean of the first two data points in the series as recommended by Lewis [22].

Lines 3366–3370 calculate the values of the error, e', represented in the program by E, and the starting values of D_1 and R_1 represented by D(1) and R(1).

A run length counter, RM, is started at line 3372 and this is an essential count for later use in Part D (lines 3432–3442), where the new process mean, M, has to be estimated before the scan can continue.

Once D_1 and R_1 are known, then the $Min(D_{N-1}, a)$ term can be executed at each new estimate of D and R within the loop, 3376–3402, at lines 3388 and 3390.

Lines 3392 and 3394 test the current and immediately preceeding values of D and R to see if they are negative. If they are, then a change in the process mean has been detected, the U or the D flag is set to one as appropriate and then line 3400 directs the program out of the loop to the Part D subroutine at line 3422. If, however, there is no such change detected then lines 3406 and 3408 check the condition of FLAG before incrementing the scan counter, C, at line 3410, and decrementing the V-mask 'a' dimension, A, by DECA. Line 3416 provides a few seconds' delay for the user to appraise the monitor display of the R values that were less than A. The next scan is then commenced by returning to line 3356. When the user has become familiar with and confident of the performance of this program, this delay loop and lines 3396 and 3398 can be deleted.

Part D is devoted to the calculations and adjustments that have to be made once two successive negative values of R() or D() have been detected. A new process mean has to be calculated, the U and C flags reset, and the run counter, RM, reset. Furthermore if there was an upward change detected in the vicinity of data items 19 to 22 then the program sets FLAG to 1 to indicate that the artificial step in the data has been detected by the current V-mask. The values of R(I) and

R(I−1) or D(I) and D(I−1) also have to be recalculated on the basis of the new process mean, rather than retain the negative values obtained on the basis of the process mean for the previous period.

To calculate the new process mean, two accumulators, DSUM and USUM, are used for the sums of the VA() values that are associated with values of D() and R() that form a continuous series with the property that they are all less than the current value of A. As soon as a value of D() or R(), as appropriate, is encountered that is not less than the current A, the trend series of VA() values is considered to have ended (see lines 3438 and 3440). DSUM and USUM are set to zero in line 3430 and are added to as a consequence of the tests in lines 3434 and 3436. Of course, there must be a count of the number of VA() values that are accumulated, and this is done by the up series counter, UC, and the down series counter, DC, initialized in line 3430 and incremented in lines 3434 and 3436 respectively. The new process mean, M, is then calculated in line 3446 or 3448 as appropriate and this value inserted into the process mean array at M(I−1) and M(I), ie. alongside the respective VA(I−1) and VA(I) values. Lines 3452 to 3468 then recalculate the D(I−1), D(I), R(I−1) and R(I) values on the basis of the new process mean. Before returning to Part C, line 3402, the U and C flags are reset to zero and the run counter to two, line 3474, and in line 3472 the very important test is made as to whether or not the change in the process mean just detected was an increase which occurred in the crucial 19th to 22nd data item window, line 3472. If it was in that window then the FLAG is set to one, which permits the current scan to be completed but prevents a new scan from being commenced, (see line 3406), and the position of the start of the new process mean is stored in the test tab variable, TT.

PART E

This routine is a strightforward piece of BASIC code that displays a report on the monitor and on the line printer of the scan results associated with the V-mask that first detected the artificial positive step in the data starting within the window from the 19th to the 22nd data item. At line 3532 the user is given the opportunity to accept this result as being suitable for the purpose by responding 'N'. Alternatively a 'Y' response in combination with line 3534 permits rejection of the recommendation and subsequent branching to Part H where the onus is placed on the user to make an alternative selection for '*a*'. Line 3536 permits continuation of the scanning routine if the response is 'C'. An 'N' entry at line 3532 followed by a 'RETURN' at line 3536 permits the execution of the final lines of Part E, lines 3540 and 3542, and progression to Part F.

PART F

The first routine (lines 3554–3558), removes the artificial step from the data between the 21st and 40th items. Lines 3566 and 3568 reopen the QC file and rewrite the header with the accepted value of A now being written in as the variable 'VMA'. The loop 3572–3576 then writes the batch dates and QC data back into the file and the program halts at line 3586.

PART H

This part of the program is entered only if the user has decided at line 3532 that the value of A arrived at by the program is too sensitive or otherwise unacceptable. In this situation the program presents the user with a complete list of all the values of A tested so far, loop 3680–3686, and then requests the user to input a final choice at line 3690. FLAG is then set to two at line 3692. The process is then returned to line 3356 in Part C to complete one final scan using this value of A, then branching at line 3408, to line 3504 in Part E, so that the user is provided with a hardcopy and softcopy report on the performance of that selection, before proceeding automatically to write this value of A as the 'VMA' into the QC file. This sequence of events is determined by the jump made at line 3530–3540, which rules out any further 'debate' with the program, and so the execution of Part F continues uninterrupted.

3.4.1 Listing of VMASKA

```
3200 'VMASKA
3202 'PROGRAM FOR DETERMINING THE VALUE OF THE V–"MASK 'A'
3204 'APPROPRIATE TO THE QC DATA TO BE MONITORED
3206 '
3208 'COMPUTERIZED QUALITY CONTROL
3210 'T F HARTLEY
3212 'PUBLISHED BY ELLIS HORWOOD, CHICHESTER, 1986
3214 '
3216 FOR I = 1 TO 5 : PRINT : NEXT I
3218 '
3220 WIDTH LPRINT 100
3222 '
3224 PRINT STRING$(60, "*")
3226 PRINT " V–MASK 'A' SELECTION PROGRAM"
3228 PRINT
3230 PRINT "THIS PROGRAM OPERATES UPON A CANDIDATE QC FILE THAT"
3232 PRINT "MUST CONTAIN 40 QC VALUES WHICH THE USER CONSIDERS"
3234 PRINT "ARE ALL NORMALLY DISTRIBUTED ABOUT THE TARGET MEAN"
3236 PRINT "FOR THE QC MATERIAL" : PRINT
3238 INPUT "NAME OF THE QC FILE .. .DAT IS NOT REQUIRED ";N$ : PRINT
```

```
3240 N$ = N$ + ".DAT"
3242 DIM DA$(100), VA(100), D(100), R(100), M(100), A$(41)
3244 '
3246 'PART A .......................................................
3248 '
3250 'READ IN DATA FROM THE NOMINATED QC FILE
3252 '
3254 OPEN "I", #1, N$
3256 INPUT#1, ID$, TM, SD, TRA, VMA, NF, FM
3258 PRINT "QC FILE IS "; ID$
3260 PRINT "TARGET QC MEAN IS "; TM
3262 PRINT "TARGET SD IS "; SD
3264 PRINT "TRIGG TRACKING ALPHA = "; TRA
3266 PRINT "V−MASK A = "; VMA
3268 PRINT "FILE CONTAINS "; NF; " ENTRIES"
3270 IF NF <40 THEN PRINT "INADEQUATE NUMBER OF DATA POINTS FOR "; :
     PRINT "THIS PROGRAM TO RUN !!!!!!!!" : STOP
3272 PRINT "FILING MARKER IS AT "; FM
3274 '
3276 FOR I = 1 TO 100
3278 INPUT#1, DA$(I), VA(I)
3280 NEXT I
3282 '
3284 CLOSE
3286 '
3288 'PART B .......................................................
3290 '
3292 'DETERMINE MEAN AND SD OF THE FILE DATA
3294 '
3296 SXX = 0 : SX = 0
3298 '
3300 FOR I = 1 TO 40
3302 SX = VA(I) + SX
3304 SXX = (VA(I)) ↑ 2 + SXX
3306 NEXT I
3308 '
3310 DSD = SQR( SXX/40 " (SX/40) ↑ 2 )
3312 MFD = SX/40
3314 PRINT
3316 PRINT "MEAN AND SD OF THE FILE DATA = "; MFD; " "; DSD
3318 PRINT
3320 '
3322 'PART C .......................................................
3324 '
3326 'OPTIMUM V−MASK A DIMENSION SEARCH ROUTINE
3328 '
3330 PRINT "SIZE OF THE STEP INCREASE IN THE RUNNING MEAN THAT"; :
     INPUT " YOU WISH TO DETECT AS A %AGE "; ST
3332 '
3334 FOR I = 21 TO 40
3336 VA(I) = VA(I) * (ST + 100) / 100
3338 NEXT I
3340 '
3342 GOSUB 3590
3344 '
3346 '
3348 C = 1 : B = 2 : U = 0 : D = 0 : FLAG = 0
3350 DECA = 3*DSD / 100
3352 A = 3 * DSD
3354 '
3356 PRINT : PRINT STRING$(60, "*")
```

```
3358 PRINT "SCAN NUMBER = "; C; " V−MASK A SET AT "; A
3360 PRINT "CANDIDATE INCREASES = R(I) VALUES < A :−"
3362 M = ( VA(1) + VA(2) ) / 2
3364 M(1) = M
3366 E = VA(1) − M
3368 D(1) = A + A/B + E
3370 R(1) = A + A/B − E
3372 RM = 1 : 'RM = RUN LENGTH COUNTER
3374 '
3376 FOR I = 2 TO 40
3378 RM = RM + 1
3380 M(I) = M
3382 E = VA(I) − M
3384 D(I) = A + A/B + E
3386 R(I) = A + A/B − E
3388 IF D(I−1) < A THEN D(I) = D(I−1) + A/B + E
3390 IF R(I−1) < A THEN R(I) = R(I−1) + A/B − E
3392 IF D(I) < 0 AND D(I−1) < 0 THEN D = 1
3394 IF R(I) < 0 AND R(I−1) < 0 THEN U = 1
3396 IF R(I) < A AND R(I) > 0 THEN PRINT I; VA(I); R(I)
3398 IF R(I) < A AND R(I) < 0 THEN PRINT I; VA(I); R(I); " ...."−E"
3400 IF D = 1 OR U = 1 THEN GOSUB 3422
3402 NEXT I
3404 '
3406 IF FLAG = 1 THEN GOTO 3480
3408 IF FLAG = 2 THEN GOTO 3504
3410 C = C + 1
3412 A = A − DECA
3414 PRINT "PAUSE NOW IF YOU WISH TO INSPECT THESE VALUES ...."
3416 FOR ET = 1 TO 600 : NEXT ET
3418 GOTO 3356
3420 '
3422 'PART D ....................................................
3424 '
3426 'TWO CHANGES IN A ROW DETECTED
3428 '
3430 DSUM = 0 : USUM = 0 : UC = 0 : DC = 0 : 'UC & DC = UP & DOWN
        COUNTERS
3432 FOR J = I TO I − RM + 1 STEP −1
3434 IF D(J) < A THEN DSUM = DSUM + VA(J) : DC = DC + 1
3436 IF R(J) < A THEN USUM = USUM + VA(J) : UC = UC + 1
3438 IF D(J) > A AND D = 1 THEN GOTO 3446 : 'END OF DOWN RUN
3440 IF R(J) > A AND U = 1 THEN GOTO 3446 : 'END OF UP RUN
3442 NEXT J
3444 '
3446 IF D = 1 THEN M = DSUM / DC
3448 IF U = 1 THEN M = USUM / UC
3450 M(I−1) = M : M(I) = M : T = I − 1 : 'T = CHANGE START INDEX
        MARKER
3452 'RECALCULATE D(I−1), R(I−1), D(I) AND R(I) FOR NEW MEAN, M
3454 E = VA(I−1) − M
3456 D(I−1) = A + A/B + E
3458 R(I−1) = A + A/B − E
3460 '
3462 E = VA(I) − M
3464 D(I) = A + A/B + E
3466 IF D(I−1) < A THEN D(I) = D(I−1) + A/B + E
3468 R(I) = A + A/B − E
3470 IF R(I−1) < A THEN R(I) = R(I−1) + A/B − E
3472 IF T >= 19 AND T <= 22 AND U = 1 THEN FLAG = 1 : TT = T
3474 U = 0 : D = 0 : RM = 2
```

```
3476 RETURN
3478 '
3480 'PART E ....................................................
3482 '
3484 'CHANGE DETECTED BETWEEN DATA ITEM 19 TO 22, FLAG = 1 POINTS
     HERE
3486 '
3488 PRINT : PRINT STRING$(60, "−")
3490 LPRINT : LPRINT STRING$(60, "−")
3492 PRINT "CHANGE DETECTED STARTING AT DATA ITEM #"; TT
3494 LPRINT "CHANGE DETECTED STARTING AT DATA ITEM #"; TT
3496 LPRINT "V−MASK A = "; A
3498 PRINT "V−MASK A = "; A
3500 PRINT "SCAN NUMBER = "; C : LPRINT "SCAN NUMBER = "; C
3502 '
3504 'PRINT OUT RESULTS SUMMARY, FLAG = 2 POINTS HERE
3506 '
3508 LPRINT : LPRINT
3510 PRINT "#"; TAB(6) "VA(#)"; TAB(16) "MEAN"; TAB(30) "#";
     TAB(38) "VA(#)"; TAB(55) "MEAN"
3512 LPRINT "#"; TAB(6) "VA(#)"; TAB(16) "MEAN"; TAB(30) "#";
     TAB(38) "VA(#)"; TAB(55) "MEAN"
3514 '
3516 FOR I = 1 TO 20
3518 J = I + 20
3520 PRINT I; : LPRINT I;
3522 PRINT TAB(6) VA(I); TAB(16) M(I); TAB(30) J; TAB(38) VA(J);
     TAB(55) M(J)
3524 LPRINT TAB(6) VA(I); TAB(16) M(I); TAB(30) J; TAB(38) VA(J);
     TAB(55) M(J)
3526 NEXT I
3528 '
3530 IF FLAG = 2 THEN GOTO 3540
3532 PRINT : INPUT "IS THIS TOO SENSITIVE .. Y/N "; Q$
3534 IF Q$ = "Y" THEN GOTO 3666
3536 IF Q$ = "N" THEN INPUT "CONT SEARCH OR OK ..... C / RETURN ";Q$
3538 IF Q$ = "C" THEN FLAG = 0 : GOTO 3410
3540 PRINT "THIS VALUE OF V−MASK A WILL NOW BE WRITTEN INTO THE "; :
     PRINT "QC FILE"
3542 VMA = A
3544 '
3546 'PART F ....................................................
3548 '
3550 'WRITE VALUE OF A INTO THE QC FILE HEADER
3552 '
3554 FOR I = 21 TO 40
3556 VA(I) = VA(I) * 100 / (100 + ST)
3558 NEXT I
3560 '
3562 PRINT "VALUE OF V−MASK A WRITTEN INTO FILE HEADER = "; A
3564 '
3566 OPEN "O", #1, N$
3568 WRITE#1, ID$, TM, SD, TRA, VMA, NF, FM
3570 '
3572 FOR I = 1 TO 100
3574 WRITE#1, DA$(I), VA(I)
3576 NEXT I
3578 '
3580 CLOSE
3582 '
3584 PRINT : PRINT "............. DONE ................." : PRINT
```

```
3586 STOP
3588 '
3590 ' PART G ..............................................
3592 '
3594 'GRAPH PLOTTING ROUTINE
3596 '
3598 PRINT : PRINT "PLOTTING GRAPH NOW .... " : PRINT
3600 A$(1) = " " + STRING$(81, "|")
3602 A$(41) = A$(1)
3604 FOR I = 2 TO 40
3606 A$(I) = "−" + STRING$(40, " ") + "|"+ STRING$(40, " ") + "−"
3608 NEXT I
3610 '
3612 MINY = MFD − 3.15 * DSD
3614 MAXY = (ST + 100) * MFD / 100 + 3.15 * DSD
3616 YD = MAXY − MINY
3618 '
3620 LPRINT : LPRINT : LPRINT TAB(18) " ";
3622 LPRINT "QC DATA FROM "; N$;" PLOTTED WITH THE "; ST; :
     LPRINT "% STEP FROM #21 TO #40"
3624 LPRINT TAB(19) "MEAN OF DATA WITHOUT THE STEP AT #21 − #40 = ";
     MFD;
3626 LPRINT " SD = "; DSD : LPRINT
3628 LPRINT TAB(19) "MINY = "; MINY; " MAX Y = "; MAXY;
     " Y INTERVALS = "; YD/40 : LPRINT
3630 '
3632 FOR J = 1 TO 40
3634 X1 = J * 2 + 1
3636 Y1 = 41 − INT(( 40 * (VA(J) − MINY) / YD) + 0.5)
3638 P$ = A$(Y1)
3640 X2 = X1 + 2
3642 X3 = 84 − X2
3644 T$ = MID$(P$, 1, X1) + "*" + MID$(P$, X2, X3)
3646 A$(Y1) = T$
3648 NEXT J
3650 '
3652 FOR I = 1 TO 41
3654 LPRINT TAB(10) A$(I)
3656 NEXT I
3658 INPUT "COLLECT QC PLOT FROM PRINTER THEN PRESS RETURN "; Q$
3660 PRINT
3662 RETURN
3664 '
3666 'PART H .......................................................
3668 '
3670 'USER WISHES TO SELECT OWN VALUE OF A
3672 '
3674 PRINT : PRINT "VALUES OF A TESTED SO FAR :−"
3676 A = 3*DSD
3678 '
3680 FOR J = 1 TO C
3682 PRINT "# "; J; TAB(12) A
3684 A = A − DECA
3686 NEXT J
3688 '
3690 INPUT "YOUR SELECTION OF A "; A
3692 FLAG = 2
3694 GOTO 3356
3696 '
```

3.4.2 Examples of the performance of the V-mask scanning routine

We established three data files to test the performance of the V-mask scanning program, VMASKA. The first file was based upon some data used by Lewis [22] on the diameters of shafts produced by an automatic lathe which was known to have an intermittent fault in the tool-feed mechanism. Lewis reported an underlying process mean of 11.36 cm for the first thirteen points, an increased process mean of 11.39 cm from the 14th to the 13th, and a return to the original 11.36 cm from the 31st to the 42nd. The data had been transposed by subtracting 11 from each and multiplying by 100. We rearranged the transposed data slightly to obtain a complete run of twenty data points before the upward shift in the process mean. In addition we introduced a new first point so that the mean of the first two points, which determine the starting value of the process mean used by the program, was exactly 36. The data are listed in Table 3.1.

These data were processed by VMASKA without introducing a percentage step increase from the 21st data item onwards, as we wished to attempt to detect the step identified by Lewis. During the 66th scan the program detected an increase starting at the 22nd data item. At that point the V-mask 'a' dimension was equal to 4.5596. The process mean for the first 21 data points was 36, and 39.8 for the remaining 19th points. We considered that these results were in acceptable agreement with the published process mean results of 36 and 39 respectively. The difference in the latter estimate was probably related to the different way in which the VMASKA program decremented the 'a' parameter. The plot of these data along with the process means are shown in Fig. 3.4; the points classified with the two process means are shown boxed in.

The second set of data used to test the performance of the VMASKA program was completely artificial and derived as follows from the random data table, (Table A1 in Appendix A). An arbitary mean of 10 and an SD of 1 were selected. The SD was then multiplied by each number in the first column of Table A1 plus the two at the top of the second column to obtain the forty random deviations required for the simulation. The sign of the deviation was then assigned according to whether the number with the corresponding position in the tenth column of Table A2 was odd or even. (If it was even then the deviation was assigned 'plus'). The procedure for the first two data items illustrates the approach:

Table 3.1 — Transposed shaft diameter data of rearranged for processing by the V-mask scanning program, VMASKA.

File No.	Value mm×100	File No.	Value mm×100
1	38.3	21	38.8
2	33.7	22	43.1
3	37.8	23	37.5
4	38.8	24	37.6
5	33.2	25	46.0
6	40.2	26	27.5
7	37.2	27	43.8
8	38.2	28	42.7
9	29.2	29	34.3
10	40.7	30	36.8
11	36.8	31	39.7
12	36.1	32	40.3
13	39.5	33	40.5
14	32.4	34	33.5
15	43.3	35	43.2
16	32.2	36	43.5
17	30.0	37	39.5
18	36.1	38	42.5
19	36.4	39	30.4
20	38.2	40	41.0

Mean = 10, deviation = 0.3. 1st item from Table A2 = 8948, hence sign for the deviation = + and 1st data item = 10.3

Mean = 10, deviation = 0.2. 2nd item from Table A2 = 2495, hence sign for the deviation = − and 2nd data item = 9.8.

The complete simulated data set is shown in Table 3.2 which has an actual mean of 9.81, an SD of 0.851 and a mean ±2 SDs range of 8.1–11.5.

These data were processed by the VMASKA program with a 10% step increase from the 21st to 40th data item. This step was detected in the 63rd scan, as starting at the 21st data item, while the V-mask 'a' dimension was 0.9575. From Fig. 3.5, it is appparent that before detecting the shift at the 21st item it made the following classifications

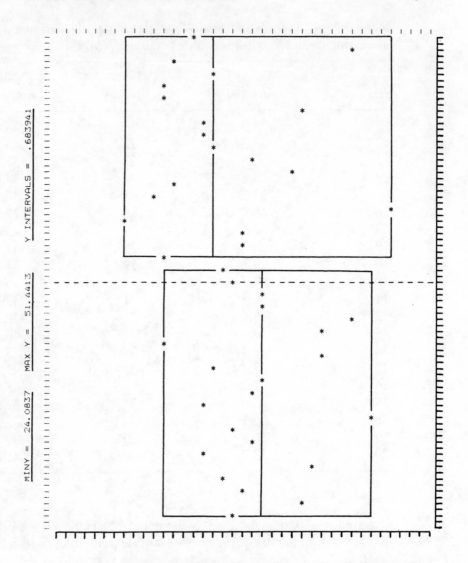

Fig. 3.4 — The performance of the VMASKA program with the shaft diameter data.

of the data:

DATA ITEMS	PROCESS MEAN
1 to 10	10.05
11 and 12	8.85
13 and 14	10.50
15 to 20	9.60

These suggested that the transients in the data, 8.5 at data item 11 and

Table 3.2 — Simulated Data

Item No.	Deviation	Data	Item No.	Deviation	Data
1	−0.3	10.3	21	−0.7	9.3
2	−0.2	9.8	22	−0.4	9.6
3	−2.0	8.0	23	−1.5	8.5
4	+0.6	10.6	24	−0.9	9.1
5	+0.1	10.1	25	+0	10.0
6	+0.3	10.3	26	−0.6	9.4
7	−0.8	9.2	27	−0.8	9.2
8	−0.8	9.2	28	+2.1	12.1
9	+0.2	10.2	29	+0.5	10.5
10	+0.7	10.7	30	−0.3	9.7
11	−1.5	8.5	31	+0.3	10.3
12	−0.8	9.2	32	−1.1	8.9
13	+1.2	11.2	33	−1.3	8.7
14	−0.2	9.8	34	+0.3	10.3
15	−1.2	8.8	35	−0.3	9.7
16	+0.2	10.2	36	+1.1	11.1
17	−0	10.0	37	−1.5	8.5
18	+0.5	10.5	38	+0	10.0
19	+0.3	10.3	39	−0.5	9.5
20	+0.3	10.3	40	+0.8	10.8

11.2 at data item 13, both tended to become associated with the next data point. A similar effect was noted for the '10% stepped up' value of 13.31 at data item 28. This was not considered to be a negative feature of the technique and as will be pointed out later it serves to augment the usefulness of the Trigg tracking signal. At this point suffice it to say that this aspect serves to draw the analyst's attention to such events; he or she should then take time to check that the transient did actually resolve itself satisfactorily within the next two or three batches. The program's estimate of the process mean from the 21st item onwards was 10.30 which was in fact 7.3% higher than the process mean reported for the items 15–20. The arithmetic mean of the first 20 data items was 9.86 and that of the second 20 was 10.74. The arithmetic mean of item 21–40, expressed as a percentage of the mean obtained for the first 20 items, was 108.9%. Hence the V-mask detection and estimate of a 7.3% increase was considered to be more than reasonable.

The third data set was based on a series of lead-in-blood quality-control results and these are shown in Table 3.3. The program took some time to settle on a realistic process mean value, as can be seen from Fig. 3.6, and did not achieve it until the ninth data item. The 10% step increment introduced into the data by the program was detected during the 32nd scan by a V-mask with 'a' = 0.1674. This V-mask reported the start of the increment as coinciding with the 22nd

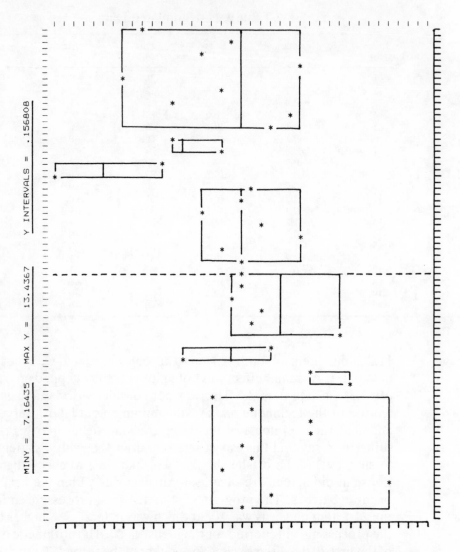

Fig. 3.5 — The simulated QC data with an artificial 10% step increment added to all
values between items 21 and 40. The V-mask 'a' dimension was 0.9575.

data item. The process mean reported between the 9th and 21st data
item was 2.433 μmol/l. and that for the remainder of the subsequent
data points was 2.574 μmol/l. which was equivalent to a 5.8% step
increase. The arithmetic mean of the first 20 points was 2.432 μmol/l.
and that of the second 20 to which the 10% increment had been added
was 2.612 μmol/l. The increment as a percentage of the first arithmetic mean was therefore +7.4%, so the V-mask's detection and
estimate of a +5.8% shift was more than acceptable.

Table 3.3 — Data included in the quality control file,
LEADQC. DAT. Units = μmol/l.

No.				
1	2.57	2.38	2.39	2.40
2	2.55	2.43	2.36	2.40
3	2.49	2.46	2.27	2.44
4	2.35	2.51	2.33	2.37
5	2.51	2.36	2.33	2.35
6	2.39	2.37	2.53	2.45
7	2.52	2.39	2.31	2.28
8	2.46	2.34	2.31	2.59
9	2.42	2.33	2.29	2.31
10	2.38	2.43	2.40	2.38

We have found the VMASKA program to be an effective tool in selecting values for the cusum V-mask '*a*' parameter. Once experience is gained with a particular assay and quality control material the value of '*a*' can be adjusted by the user for fine tuning.

3.5   THE TRIGG TRACKING SIGNAL

The tracking signal proposed by D. W. Trigg [25] is an extension of the earlier work by R. G. Brown [31], in which a tracking signal was obtained by dividing the cusum by the mean absolute deviation, MAD:

Brown's tracking signal = Cusum/MAD

where MAD = $(1 - \alpha)$. previous MAD + $\alpha \times$ latest absolute error and α is a smoothing constant with a value less than one.

Trigg found two disadvantages of this formula, one being that if the tracking signal indicated that a process had gone out of control and the problem was actually rectified, the tracking signal did not fall back into limits without user intervention. Secondly, if the process was in excellent control, the MAD would tend to fall towards a smaller and smaller value while the cusum would continue to increase. Under these circumstances the value of the tracking signal would increase and eventually become significant. Trigg's solution was to use a smoothed estimate of the cusum, or *smoothed error*,

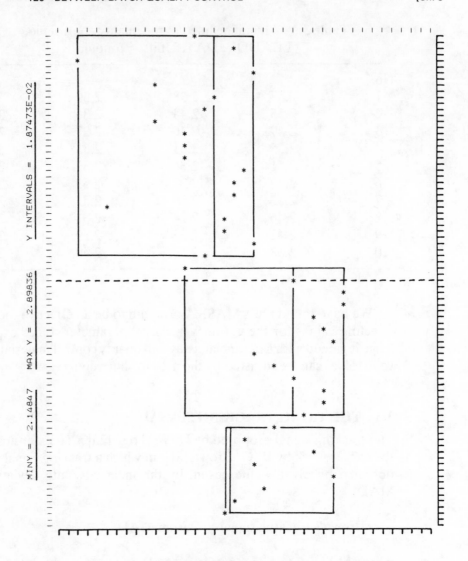

Fig. 3.6 — The blood-lead quality control data. The V-mask '*a*' dimension was 0.1674. Y-Axis units = μmol/l.

which restrained the signal to limits of plus or minus one:

$$\text{Smoothed Error} = (1 - \alpha) \times \text{previous smoothed error} + \alpha \times \text{latest error}$$

The Trigg tracking signal has been used in a patient-monitoring system [26] and we used the same algorithm and notation in our program. The program starts by initiating the following variables:

V_{ave}=current assigned value of the QC material

Let V_{ave}=the exponentially weighted average for the previous sampling period

$\qquad = U_{t-1}$

Let the previous error in the prediction $= \bar{e}_{t-1} = V_{ave}/100$

Let the previous MAD $= \text{MAD}_{t-1} = V_{ave}/100$

Let d_t = the current measured value of the QC material.

The main subroutine is as follows:

$\ddagger U_t = \alpha \times d_t + (1 - \alpha) \times U_{t-1}$

$\qquad \bar{e}_t = \alpha\,(d_t - U_{t-1}) + (1 - \alpha) \times \bar{e}_{t-1}$

$\qquad \text{MAD}_t = \alpha\,|(d_t - U_{t-1})| + (1 - \alpha) \times \text{MAD}_{t-1}$

$\qquad T_t$ = Trigg tracking signal = $\bar{e}_t MAD_t$

'Moving' estimate of the SD $= 1.2533 \left(\dfrac{2 - \alpha}{2}\right)^{1/2} \text{MAD}_t$

This final equation is based on one given by Cembrowski et al [32]. Before the next estimate of the tracking signal can be made, the current values of U_t, $\bar{e}_t$ and MAD_t have to be assigned to U_{t-1}, $\bar{e}_{t-1}$ and MAD_{t-1} and the calculations recommenced from $\ddagger$ using the latest measured value for the QC material as d_t.

The selection of different values for α provides variable degrees of smoothing. If $\alpha = 0.1$, the calculations are based on exponentially weighted means of a series of nineteen observations. We usually assign a value of 0.2 to α which means that our tracking signal is under the exponentially weighted influence of the previous nine values for the quality control material. The value of α can be tailored exactly to the user's requirements because n, the number of observations embraced by the exponential weighting calculations, can be selected by using the following relationship:

$$n = (2/\alpha) - 1$$

The values of T are normally distributed about a mean of zero, so confidence limits are available to assist with the interpretation of the results for T. For an α of 0.2 the 95% confidence limits are ± 0.58. Table 3.4 presents a selected series of values for α and the corres-

Table 3.4 — Values for α and the corresponding number of preceeding observations that influence the smoothing process.

α	Number of Observations	α	Number of Observations
0.0198	100	0.2222	8
0.0263	75	0.2500	7
0.0392	50	0.2857	6
0.0952	20	0.3333	5
0.1000	19	0.4000	4
0.1818	10	0.5000	3
0.2000	9	0.6667	2

ponding number of observations embraced by the smoothing process.

Batty [33] addressed the problem of assigning conventional statistical confidence limits to the Trigg tracking signal, and he derived appropriate tables. Following on from this, Cembrowski *et al.* [32] examined the calculation and simulation techniques used to derive these limits. We have followed their suggestions and chosen to use the results from the simulation experiments for values of α greater than 0.1, and the calculated limits for α values equal to or less than 0.1 to interpret the Trigg tracking signal produced from our between batch quality-control program called REPORT. A list of useful confidence limits for various values of α are shown in Table 3.5.

Table 3.5 — Probability of the Trigg tracking signal exceeding these values

α	$p=0.10$	$p=0.05$	$p=0.001$
0.100	0.34	0.41	0.54
0.200	0.50	0.58	0.71
0.300	0.63	0.71	0.82
0.400	0.72	0.80	0.92
0.500	0.82	0.88	0.94

3.6 A PROGRAM TO PRODUCE A SUMMARY REPORT ON BETWEEN-BATCH QUALITY-CONTROL DATA

This section describes a program called REPORT that produces a summary of the data in a nominated QC file. The reports include a standard normal deviate plot alongside of which are the batch dates, the Trigg tracking signal and the process mean determined by the cusum V-mask. The hardcopy has two additional columns which contain the actual QC value and the moving estimate of the standard deviation.

The array variables used in this program are declared in the dimension statement in line 3786. Their assignments are as follows:

DA$(100) : batch dates
VA(100)   : QC values
SND(100) : standard normal deviates
TS(100)   : Trigg tracking signals
MPM(100) : moving process means determined by the cusum V-mask
MSD(100) : moving estimates of the standard deviation
UT(2)   : exponentially weighted averages for the current and previous sampling periods
EB(2)   : current and previous errors in the predictions
MAD(2)   : current and previous estimates of the mean absolute deviations
D(100)   : the cusum V-mask downward trend in the process-mean test parameter
R(100)   : the cusum V-mask upward trend in the process-mean test parameter

PART A
This section (lines 3790–3832) reads in the batch dates and QC data from the nominated QC file, N$. It is almost identical to Part A of the VMASKA program and therefore requires no further description.

PART B
This section (lines 3834–3854), performs the simple standard normal deviate calculations and stores the results in the SND() array.

Standard Normal Deviate = (QC Value − Target Mean)/Target SD

PART C
This is a new piece of code that implements the Trigg tracking signal algorithm described by Hope *et al.* [26].

At lines 3868–3872, the initial values, (t=1), of the variables UT, EB and the MAD are assigned as being equal to the target mean, 1% of the target mean and 10% of the target mean respectively.

Line 3876 checks whether the file contains 100 entries. If it does, the data-reading sequence starts from the file marker to the 100th item in the VA() array, loop 3896–3900. The remaining data from the first item in the VA() array to the (FM − 1) th item is then read by the loop 3904–3908. In this way the Trigg tracking signal is calculated, by the subroutine 3916–3934, in strict chronological order. If the file contains less than 100 entries, the loop at lines 3880–3884 is used.

The common Trigg tracking signal routine (lines 3916–3934), follows the formulae of Hope *et al.* [26], with an additional calculation of the moving estimate of the standard deviation as defined by Cembrowski *et al.*, [32]:

$$(\pi/2)^{\frac{1}{2}} [(2 - \alpha)/2]^{\frac{1}{2}} \, \text{MAD}_t$$

which for the purposes of this program we have reduced to:

$$1.2533 \, [(2 - \alpha)/2]^{\frac{1}{2}} \, \text{MAD}_t$$

Line 3932 is important because it is here that the current values of EB, UT and MAD are assigned to 'previous' status before the next Tracking Signal is computed.

PART D

This section of the program is very similar to the code used in Parts C and D of the VMASKA program and only the differences will be highlighted. The loop 3970–3974 caters for the chronology of files with less than 100 entries and the loops 4000–4004 and 4008–4012 for files with 100 entries.

The common cusum V-mask scanning routine, lines 4020–4042, is a piece of code identical to lines 3378–4042 of the VMASKA program.

Lines 4048–4102 are identical to lines 3426–3476 in the VMASKA program except that this program, REPORT, does not require the test performed at line 3472 or the use of the variable T, the 'change start index marker', line 3450.

PART E

This is the monitor display and hard copy print-out routine. As was mentioned in the section on the QCFILER program, optimum use must be made of the space available on the monitor and the printer. At line 4114, the line width of the monitor is set to 80 columns from the default value of 72 characters. At line 4118, the line width on the printer is set to 110 columns. If a hard copy print-out is required then the flag HC is set to 1 following a 'Y' input from the user at line 4118. Line 4122 prints the title of the report.

Line 4124 prints the values of the target mean minus two target standard normal deviates, the target mean and the target mean plus two target standard normal deviates at the appropriate tab positions for the standard normal deviate plot that follows on from this. The scale for this plot is generated from S$ and the loop 4132–4140 and stored as SC$. This scale is printed by line 4144. The scale is actually in standard normal deviate units: zero + and − 2.5 SNDs, but since line 4122 has printed the corresponding values in the user's units of measurement it is possible to interpret the plot in either units.

The three loops 4146–4152, 4160–4164 and 4168–4172 are the now familiar routines for dealing with partially filled and full data files in their correct chronological order. Line 4176 is the final line to be executed by this program.

Lines 4180–4230 constitute the common display and printout routines called from within the three loops just mentioned. The monitor displays up to five screens, if there are 100 entries, 20 lines at a time in chronological order oldest to the most recent. A line counter, L, is used at line 4188, tested at line 4222 and reset at line 4224 for this purpose. Along the top of each display on the monitor are the values of the target mean minus two target SD's, the target mean, and the target mean plus two target SD's at the appropriate tab stops (line 4184). Line 4186 prints out a scale which starts at minus 2.5 target SD's, through zero, to plus 2.5 target SD's with an interval resolution of 0.1 target SD units. The code from 4190–4204 is similar to the approach used in the linear calibration graph program, line 430. The position of a point in a standardized string format, A$ as defined by line 4128, is first determined as an appropriately scaled integer value, P, line 4190, and then placed within the interim string, B$, using the MID$ function, lines 4200, 4202 or 4204. In addition a FILL$ string has been used to place a series of '−' between the centre zero and the data point to give the plot more substance so that its trend line/centre of gravity will be more appparent to the eye. Note that standard normal deviates equal to or greater than ±2.5 SND's are all plotted on the + or −2.5 SNDs margin; i.e. outliers are not distinguished by means of a separate symbol.

The batch date is transcribed into the interim string, C$, at line 4206.

Lines 4208 and 4210 have the effect of reducing the Trigg tracking signal to a five character string, e.g. "−0.96", and storing it as D$. Similarly lines 4214–4218 reduce the moving process mean, MPM, to

a six character string called E$. Line 4220 prints the strings for the
standard normal deviate plot, B$, the batch date, C$, the Trigg
Tracking Signal, D$, and the moving process mean, E$, across the 80
columns of the monitor screen. Line 4226 performs the same function
for the hard-copy report and adds the QC value. Line 4228 prints the
moving estimate of the standard deviation, MSD, to not more than
five decimal places.

3.6.1   Listing of REPORT

```
3750 'REPORT
3752 'QC REPORT PROGRAM : THIS PROGRAM PROCESSES THE DATA IN A NAMED
3754 'QC FILE AND PRODUCES A SOFTCOPY AND AN OPTIONAL HARDCOPY
3756 'REPORT OF THE STANDARD NORMAL DEVIATE PLOT, THE BATCH DATE,
3758 'THE TRIGG TRACKING SIGNAL AND THE CUSUM V–MASK DERIVED PROCESS
3760 'PROCESS MEAN
3762 '
3764 'COMPUTERIZED QUALITY CONTROL
3766 'T F HARTLEY
3768 'PUBLISHED BY ELLIS HORWOOD, ENGLAND, 1986
3770 '
3772 FOR I = 1 TO 5 : PRINT : NEXT I
3774 '
3776 PRINT STRING$(60, "*")
3778 PRINT " QC REPORT PROGRAM"
3780 PRINT STRING$(60, "*")
3782 PRINT
3784 '
3786 DIM DA$(100), VA(100), SND(100), TS(100), MPM(100), MSD(100),
     UT(2), EB(2), MAD(2), D(100), R(100)
3788 '
3790 'PART A .......................................................
3792 '
3794 'READ IN DATA FROM NAMED QC FILE
3796 '
3798 INPUT "NAME OF QC FILE TO REPORT ON .DAT NOT REQUIRED"; N$
3800 N$ = N$ + ".DAT"
3802 OPEN "I", #1, N$
3804 INPUT#1, ID$, TM, SD, TRA, VMA, NF, FM
3806 PRINT : PRINT "QC FILE IS "; ID$
3808 PRINT "TARGET MEAN IS "; TM
3810 PRINT "TARGET SD IS = "; SD
3812 PRINT "TRIGG TRACKING SIGNAL ALPHA IS "; TRA
3814 PRINT "V – MASK A IS "; VMA
3816 PRINT : PRINT "FILE CONTAINS "; NF; " ENTRIES"
3818 PRINT "FILING MARKER IS AT "; FM
3820 '
3822 FOR I = 1 TO 100
3824 INPUT#1, DA$(I), VA(I)
3826 NEXT I
3828 '
3830 CLOSE
3832 '
3834 'PART B .......................................................
3836 '
3838 'CONVERT QC DATA TO STANDARD NORMAL DEVIATE UNITS
3840 '
```

```
3842 PRINT
3844 PRINT "STANDARD NORMAL DEVIATE CALCULATIONS IN PROGRESS"
3846 '
3848 FOR I = 1 TO NF
3850 SND(I) = ( VA(I) − TM ) / SD
3852 NEXT I
3854 '
3856 'PART C ......................................................
3858 '
3860 'DETERMINE THE TRIGG TRACKING SIGNALS AND MOVING ESTIMATE
     OF THE SD
3862 '
3864 PRINT
3866 PRINT "TRIGG TRACKING SIGNAL AND MOVING SD CALCULATIONS IN ";:
     PRINT "PROGRESS"
3868 UT(1) = TM : 'UT = u subscript t−1
3870 EB(1) = TM / 100 : 'EB = e bar
3872 MAD(1) = TM / 10
3874 '
3876 IF NF = 100 THEN GOTO 3892
3878 '
3880 FOR I = 1 TO NF
3882 GOSUB 3916
3884 NEXT I
3886 '
3888 GOTO 3938 : 'PART D
3890 '
3892 'FILE CONTAINS 100 ENTRIES
3894 '
3896 FOR I = FM TO 100
3898 GOSUB 3916
3900 NEXT I
3902 '
3904 FOR I = 1 TO FM − 1
3906 GOSUB 3916
3908 NEXT I
3910 '
3912 GOTO 3938 : 'PART D
3914 '
3916 'COMMON TRIGG TRACKING SIGNAL SUBROUTINE
3918 '
3920 UT(2) = TRA * VA(I) + ( 1 − TRA ) * UT(1)
3922 E = VA(I) − UT(1)
3924 EB(2) = TRA * E + ( 1 − TRA ) * EB(1)
3926 MAD(2) = TRA * ABS(E) + ( 1− TRA ) * MAD(1)
3928 TS(I) = EB(2) / MAD(2)
3930 MSD(I) = 1.2533 * ((( 2 − TRA ) / 2 ) ↑ 0.5 ) * MAD(2)
3932 EB(1) = EB(2) : UT(1) = UT(2) : MAD(1) = MAD(2)
3934 RETURN
3936 '
3938 'PART D ......................................................
3940 '
3942 'MOVING PROCESS MEAN FROM CUSUM V − MASK SCAN
3944 '
3946 PRINT : PRINT "CUSUM V−MASK SCAN IN PROGRESS"
3948 A = VMA : B = 2 : K = A/B
3950 '
3952 IF NF = 100 THEN GOTO 3982
3954 '
3956 MPM(1) = ( VA(1) + VA(2) ) / 2
3958 M = MPM(1) : 'MPM = MOVING PROCESS MEAN
3960 E = VA(1) − M
```

```
3962 D(1) = A + K + E
3964 R(1) = A + K – E
3966 RM = 1 : 'RM = RUN LENGTH COUNTER
3968 '
3970 FOR I = 2 TO NF
3972 GOSUB 4020
3974 NEXT I
3976 '
3978 GOTO 4106 : 'PART E
3980 '
3982 'FILE CONTAINS 100 ENTRIES
3984 '
3986 MPM(FM) = ( VA(FM) + VA(FM + 1) ) / 2
3988 M = MPM(FM)
3990 E = VA(FM) – M
3992 D(FM) = A + K + E
3994 R(FM) = A + K – E
3996 RM = 1
3998 '
4000 FOR I = FM + 1 TO 100
4002 GOSUB 4020
4004 NEXT I
4006 '
4008 FOR I = 1 TO FM – 1
4010 GOSUB 4020
4012 NEXT I
4014 '
4016 GOTO 4106 : 'PART E
4018 '
4020 'COMMON V – MASK SCANNING ROUTINE
4022 '
4024 RM = RM + 1
4026 MPM(I) = M
4028 E = VA(I) – M
4030 D(I) = A + K + E
4032 R(I) = A + K – E
4034 IF D(I–1) < A THEN D(I) = D(I–1) + K + E
4036 IF R(I–1) < A THEN R(I) = R(I–1) + K – E
4038 IF D(I) < 0 AND D(I–1) < 0 THEN D = 1
4040 IF R(I) < 0 AND R(I–1) < 0 THEN U = 1
4042 IF D = 1 OR U = 1 THEN GOSUB 4048 : 'TWO CHANGES IN A ROW
4044 RETURN
4046 '
4048 'TWO CHANGES IN A ROW DETECTED
4050 '
4052 DSUM = 0 : USUM = 0 : UC = 0 : DC = 0
4054 '
4056 FOR J = I TO I – RM STEP –1
4057 IF J <= 0 THEN JJ = 100 + J ELSE JJ = J
4058 IF D(JJ) < A THEN DSUM = ·DSUM + VA(JJ) : DC = DC + 1
4060 IF R(JJ) < A THEN USUM =  USUM + VA(JJ) : UC = UC + 1
4062 IF D(JJ) > A AND D = 1 THEN GOTO 4070 : 'END OF DOWN RUN
4064 IF R(JJ) > A AND U = 1 THEN GOTO 4070 : 'END OF UP RUN
4066 NEXT J
4068 '
4070 IF D = 1 THEN M = DSUM / DC
4072 IF U = 1 THEN M = USUM / UC
4074 MPM(I–1) = M : MPM(I) = M
4076 '
4078 'RECALCULATE D(I–1), R(I–1), D(I) AND R(I) FOR THE NEW MEAN, M
4080 '
4082 E = VA(I–1) – M
```

```
4084 D(I−1) = A + K + E
4086 R(I−1) = A + K − E
4088 '
4090 E = VA(I) − M
4092 D(I) = A + K + E
4094 IF D(I−1) < A THEN D(I) = D(I−1) + K + E
4096 R(I) = A + K − E
4098 IF R(I−1) < A THEN R(I) = R(I−1) + K − E
4100 U = 0 : D = 0 : RM = 2
4102 RETURN
4104 '
4106 'PART E ...................................................
4108 '
4110 'RESULTS DISPLAY AND PRINTOUT
4112 '
4114 WIDTH 80
4116 PRINT STRING$(60, "−") : PRINT
4118 INPUT "DO YOU REQUIRE A HARDCOPY REPORT .... Y/N "; Q$
4120 IF Q$ = "Y" THEN HC = 1 : WIDTH LPRINT 110
4122 IF HC = 1 THEN LPRINT : LPRINT N$, "TARGET MEAN IS "; TM,
       "TARGET SD IS "; SD, "TRIGG TRACKING SIGNAL ALPHA = "; TRA,
       "V − MASK A = "; VMA, "FILE CONTAINS "; NF; " ENTRIES" :
       LPRINT
4124 IF HC = 1 THEN LPRINT TAB(6) (TM − 2*SD); TAB(26) TM;
       TAB(46) (TM + 2*SD)
4126 '
4128 A$ = " " + STRING$(5, " ") + ":" + STRING$(19, " ") + "|" +
       STRING$(19, " ") + ":" + STRING$(5, " ") + " "
4130 '
4132 S$ = "....:"
4134 FOR Z = 1 TO 10
4136 SC$ = SC$ + S$
4138 NEXT Z
4140 SC$ = "." + SC$
4142 '
4144 IF HC=1 THEN LPRINT SC$
4146 IF NF = 100 THEN GOTO 4160
4148 FOR I = 1 TO NF
4150 GOSUB 4180
4152 NEXT I
4154 '
4156 GOTO 4176 : 'STOP
4158 '
4160 FOR I = FM TO 100
4162 GOSUB 4180
4164 NEXT I
4166 '
4168 FOR I = 1 TO FM − 1
4170 GOSUB 4180
4172 NEXT I
4174 '
4176 STOP : '.......................................
4178 '
4180 'COMMON DISPLAY / PRINTOUT ROUTINE
4182 '
4184 IF L = 0 THEN PRINT TAB(6) (TM − 2*SD); TAB(26) TM;
       TAB(46) (TM + 2*SD)
4186 IF L = 0 THEN PRINT SC$
4188 L = L + 1
4190 P = INT( 10 * SND(I) + 27.5 )
4192 IF P < 2 THEN P = 2
4194 IF P > 52 THEN P = 52
```

```
4196 IF P<27 THEN FILL$ = "*" + STRING$( (26"P), "−")
4198 IF P>27 THEN FILL$ = STRING$( (P−28), "−") + "*"
4200 IF P<27 THEN B$ = MID$( A$, 1, P−1) + FILL$ + MID$(A$, 27, 25)
4202 IF P>27 THEN B$ = MID$(A$, 1, 27) + FILL$ + MID$(A$, P+1, 51−P)
4204 IF P=27 THEN B$ = MID$(A$, 1, 26) + "*" + MID$(A$, 28, 24)
4206 C$ = DA$(I)
4208 TS = (INT (TS(I)*100 + 0.5 ))/100
4210 D$ = STR$( TS )
4212 D$ = MID$(D$, 1, 5)
4214 MPM = ( INT( MPM(I) * 1000 + 0.5 )) / 1000
4216 E$ = STR$( MPM )
4218 E$ = MID$(E$, 1, 6)
4220 PRINT B$; TAB(54) C$; TAB(63) D$; TAB(68) E$
4222 IF L = 20 THEN INPUT "PRESS RETURN TO CONTINUE ...."; Q$
4224 IF L = 20 THEN L = 0
4226 IF HC=1 THEN LPRINT B$; TAB(54) C$; TAB(63) D$; TAB(68) E$;
     TAB(78) VA(I);
4228 IF HC = 1 THEN MSD = (INT (MSD(I) * 10000 + 0.5)) / 10000 :
     LPRINT TAB(90) MSD
4230 RETURN
4232 '
```

3.6.2 Examples of the performance of the between-batch quality control program 'REPORT'

We start this section by using REPORT to process data from two data files called TRIGGQC.DAT and QC2.DAT. The data, and some results from processing these files are shown in Table 3.6. The data were selected to demonstrate:

(i) When an alpha value of 0.2 is used, the Trigg tracking signal responds to a sustained drift within the first two or three data points.

(ii) An increase in the variance of the data does not cause the Trigg tracking signal to indicate a persistent out of limits condition. Under these circumstances the moving estimate of the standard deviation reliably signals the deterioration in the precision of the assay.

Note that the data in these files are not meant to be representative of typical QC data files; they contain simulation data designed to highlight the characteristics cited in (i) and (ii) above.

Figure 3.7 is a hard copy equivalent of the monitor display provided by REPORT from the TRIGGQC.DAT file. The first 10 data points are a baseline period to permit the Tracking Signal to settle in before challenging it with two rapidly moving positive trends between data items 11 and 16 and then 17 to 21. Data items 22 to 26 form another short baseline period before there is a transition into a phase of rapidly oscillating values with excursions from the target

Table 3.6 — Data used to illustrate the performance of the REPORT program with special reference to the Trigg tracking signal and the moving estimate of the standard deviation.

TRIGGQC.DAT File	QC2.DAT File
Target mean = 100	Target mean = 10
Actual mean = 112	Actual mean = 9.42
Target SD = 36	Target SD = 0.90
Actual SD = 35.2	Actual SD = 0.94
Trigg tracking alpha = 0.2	Trigg tracking alpha = 0.2
V-Mask 'a' = 40	V-Mask 'a' = 0.9575

No.	Data	Trigg	MPM	MSD	Data	Trigg	MPM	MSD
1	100	0.10	100	9.51	10.3	0.16	10.05	1.02
2	100	0.10	100	7.61	9.8	0.08	10.05	0.88
3	95	−0.08	100	7.28	8.0	−0.36	10.05	1.18
4	115	0.35	100	9.63	10.6	−0.08	10.05	1.18
5	100	0.26	100	8.22	10.1	−0.01	10.05	1.02
6	85	−0.21	100	10.56	10.3	0.10	10.05	0.92
7	80	−0.48	100	12.83	9.2	−0.11	10.05	0.91
8	95	−0.47	100	10.33	9.2	−0.26	10.05	0.87
9	100	−0.28	100	9.50	10.2	−0.07	10.05	0.82
10	105	0	100	9.78	10.7	0.20	10.05	0.87
11	118	0.38	100	12.66	8.5	−0.20	8.85	1.05
12	120	0.57	100	14.48	9.2	−0.30	8.85	0.95
13	140	0.75	100	19.81	11.2	0.14	10.50	1.15
14	160	0.85	160	27.19	9.8	0.11	10.50	0.94
15	180	0.91	160	35.58	8.8	−0.17	9.60	1.01
16	200	0.94	160	44.29	10.2	−0.01	9.60	0.94
17	110	0.56	160	44.17	10.0	0.05	9.60	0.80
18	130	0.46	160	37.58	10.5	0.25	9.60	0.80
19	150	0.51	160	33.03	10.3	0.33	9.60	0.73
20	170	0.62	160	33.55	10.3	0.40	9.60	0.65
21	190	0.72	160	37.30	9.15	−0.02	9.60	0.74
22	100	0.20	160	42.87	9.30	−0.21	9.60	0.73
23	100	−0.08	100	44.73	8.05	−0.54	8.28	1.00
24	100	−0.25	100	44.12	8.50	−0.64	8.28	1.02
25	100	−0.37	100	41.97	9.25	−0.63	8.28	0.82
26	100	−0.46	100	38.92	8.50	−0.71	8.28	0.83
27	105	−0.51	100	34.22	8.30	−0.78	8.28	0.85
28	95	−0.58	100	32.22	11.20	0.01	10.40	1.22
29	110	−0.59	100	26.08	9.60	0.06	10.40	1.03
30	90	−0.67	100	25.87	8.80	−0.11	9.27	0.97
31	110	−0.61	100	21.45	9.40	−0.08	9.27	0.80
32	80	−0.72	100	23.69	8.00	−0.38	9.27	0.95
33	120	−0.40	100	23.24	7.80	−0.56	8.40	1.06
34	70	−0.59	100	27.05	9.40	−0.33	8.40	0.99
35	130	−0.18	100	29.14	8.80	−0.36	8.40	0.82
36	60	−0.44	100	33.96	10.20	0.08	8.40	0.97
37	140	−0.04	100	37.67	7.60	−0.27	8.40	1.15
38	50	−0.33	100	43.14	9.10	−0.20	8.40	0.98
39	150	0.04	100	47.89	8.60	−0.26	8.40	0.85
40	40	−0.26	100	53.76	9.90	0.08	8.40	0.93

Trigg = Trigg tracking signal.
MPM = Moving process mean.
MSD = Moving estimate of the SD.

mean getting progressively larger in each direction.

Figure 3.8 presents plots of the moving estimate of the SD and the

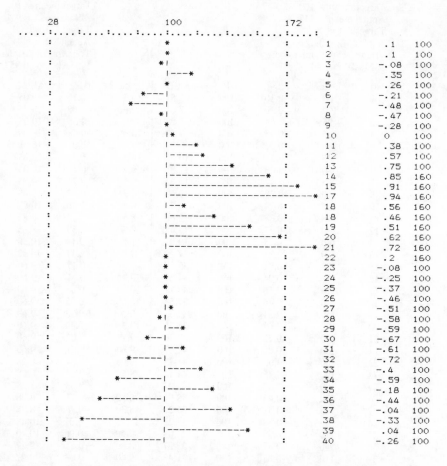

```
                    Target mean = 100     Target SD = 36
                    Trigg Tracking Signal Alpha = 0.2
                    V-Mask 'a' = 40


          28                          100                   172
     .....:....:....:....:....:....:....:....:....:....:....:
          :                           *                     :      1         .1    100
          :                           *                     :      2         .1    100
          :                          *I                     :      3       -.08    100
          :                          I---*                  :      4        .35    100
          :                           *                     :      5        .26    100
          :                      *---I                      :      6       -.21    100
          :                    *-----I                      :      7       -.48    100
          :                          *I                     :      8       -.47    100
          :                          *                      :      9       -.28    100
          :                        I*                       :     10         0     100
          :                        I----*                   :     11        .38    100
          :                        I-----*                  :     12        .57    100
          :                        I----------*             :     13        .75    100
          :                        I----------------*  :          14        .85    160
          :                        I--------------------*        :  15        .91    160
          :                        I-----------------------*     :  17        .94    160
          :                        I--*                   :     18        .56    160
          :                        I-------*              :     18        .46    160
          :                        I-------------*        :     19        .51    160
          :                        I----------------*:         20        .62    160
          :                        I----------------------*    :  21        .72    160
          :                          *              :          22         .2    160
          :                          *              :          23       -.08    100
          :                          *              :          24       -.25    100
          :                          *              :          25       -.37    100
          :                          *              :          26       -.46    100
          :                         I*              :          27       -.51    100
          :                        *I               :          28       -.58    100
          :                        I--*             :          29       -.59    100
          :                      *--I               :          30       -.67    100
          :                        I--*             :          31       -.61    100
          :                     *-----I             :          32       -.72    100
          :                     I------*            :          33        -.4    100
          :                  *--------I             :          34       -.59    100
          :                        I-------*        :          35       -.18    100
          :                *-----------I            :          36       -.44    100
          :                     I----------*        :          37       -.04    100
          :              *-------------I            :          38       -.33    100
          :                     I------------*      :          39        .04    100
          :        *-----------------I              :          40       -.26    100
```

Fig. 3.7 — Results of processing the data in TRIGGQC.DAT with the REPORT program.

Trigg tracking signal, the data for which are shown in the third and fifth columns of Table 3.6. The plot of the Trigg tracking signal illustrates how it rapidly moved beyond the +0.58 limit between the 12th and 13th data points. The latter point is the third point in the first trend series so the tracking signal had been particularly responsive in detecting this trend. The second trend series starts from the 17th point and by the 20th point the tracking signal was beyond the + 0.58 limit. Notice also how rapidly the tracking signal fell back within limits at the drop coincident with the 17th point and again at the 22nd

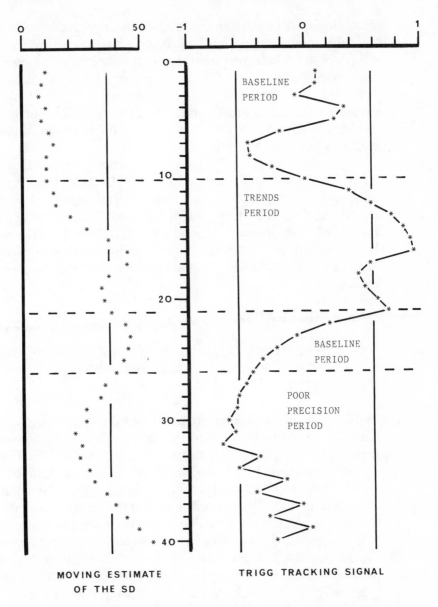

MOVING ESTIMATE
OF THE SD

TRIGG TRACKING SIGNAL

Fig. 3.8 — Plots of the moving estimate of the SD and the Trigg Tracking Signal from data generated by the REPORT program from the TRIGGQC.DAT file.

point. The tracking signal appeared to be tending to regard the unrealistically flat period of baseline between the 22nd and 26th points as a negative trend. However, the high variance region was encountered before it could clarify this and by the 33rd point the tracking signal was moving back into the acceptance limits and becoming more positive again. In addition it appeared to be 'hunting' the Trigg tracking signal zero, illustrating how insensitive it is to wild swings of the points, provided they occur about a stable mean. The moving estimate of the SD, however, is sensitive to this type of behaviour, and from the 31st point onwards a steady climb in this estimate becomes clear. It is interesting to note that the actual SD of the 36th to 40th points is in fact 52.6 and the moving estimate is in excellent agreement at 53.8.

The data in the QC2.DAT file are less contrived than those in the TRIGGQC.DAT file. They are based on QC1.DAT (shown in Table 3.2), but a trend ending in a final 'negative bias plateau' was introduced from the 21st data item onwards. This was done by using the QCFILER program to load the QC1.DAT file into memory before halting the program at line 3082, as previously described in Section 3.3. The following lines were then executed in the immediate mode:

```
N$ = "QC2.DAT"
KK = 0.15
FOR J=21 TO 26 : VA(J)=VA(J)−KK : KK=KK+0.15 : NEXT J
FOR J=27 TO 40 : VA(J)=VA(J)−0.9 : NEXT J
GOTO 3118
```

In this way we introduced a progressive negative trend between data items 21 and 26 inclusive, and thereafter subtracted a constant 0.9 from all the remaining data items in the file. Fig. 3.9 illustrates the monitor report and Fig. 3.10 a plot of the moving estimates of the SD and the Trigg tracking signals as shown in the 9th and 7th columns of Table 3.6. Figure 3.10 demonstrates that the moving estimate of the SD did not develop a discernable trend either before or after the 21st to 27th data items. The Trigg tracking signal, however, had detected the trend by the 24th data item, which was actually the fourth item in the trend series, but because no action was taken to correct this bias it eventually moved back into the acceptance limits of ±0.58. The performance characteristic of the Trigg tracking signal was mentioned in the opening paragraphs of this chapter. For completeness we have included the results of processing the data in the QC1.DAT file (Table 3.2) in Fig. 3.11. From this it can be seen that the Trigg tracking signals obtained for the 'parent' data set were all within acceptance limits. In addition, the 3rd and 28th data points, which were the two outliers in the set, were not associated with the largest tracking signals, as expected. They were, however, associated with the largest values of the moving estimate of the SD, also as expected.

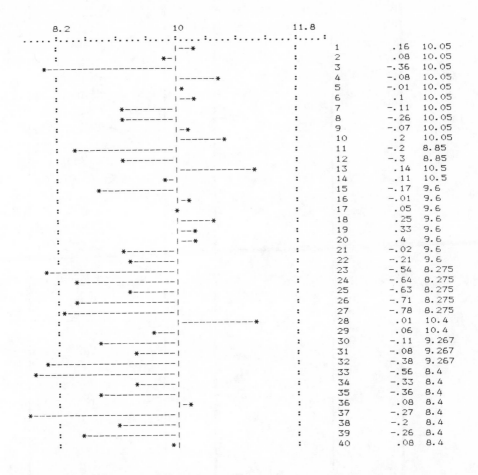

Fig. 3.9 — Results of processing the data in QC2.DAT with the REPORT program.

Finally we processed the data items in the LEADQC.DAT file, (Table 3.3), and a copy of the'monitor' report is shown in Fig. 3.12. These results were interesting because they presented Trigg tracking signal evidence to support the visual impression that there was a negative trend in these data, particularly in the region of the 23rd to 25th data points. The cusum V-mask, however, did not detect any fall, so although there was evidence of a trend it did not appear that it had approached a new process mean level. The associated moving estimates of the SD were acceptably stable and accurate.

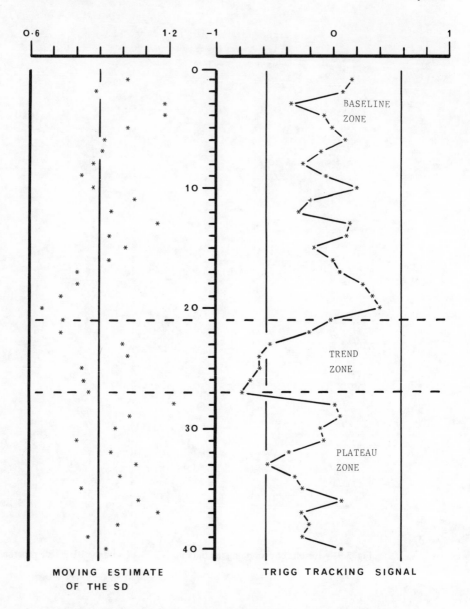

Fig. 3.10 — Plots of the moving estimate of the SD and the Trigg tracking signal from data generated by the REPORT program from the QC2.DAT file.

Apart from a brief comment in the previous paragraph, the cusum V-mask estimate of the process mean has not been mentioned in this particular discussion. The best practical illustration of its typical performance was associated with the QC2.DAT file data, shown in Table 3.7.

Target Mean = 10   Target SD = 0.9
Trigg Tracking Signal Alpha = 0.2
V-Mask 'a' = 0.9575

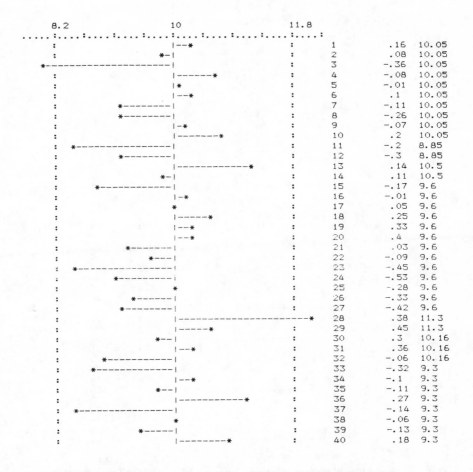

Fig. 3.11 — Results of processing the data in QC1.DAT with the REPORT program.

For this data set it demonstrated the tendency for the process mean estimate to lag behind the slower shorter-term changes, hence underestimating them, and following the faster changes closer to their extremes. This behaviour was quite apparent also in the worked examples on the VMASKA program, where transients were usually linked by the program to the next highest value on either side of the transient. Actually, this is an advantage, because while the Trigg tracking signal was steadily working itself back into the 'within limits' zone from the 33rd item onwards (because of the 'analyst's neglect' to

Target Mean = 2.43   Target SD = 0.08
Trigg Tracking Signal Alpha = 0.2
V-Mask 'a' = 0.1674

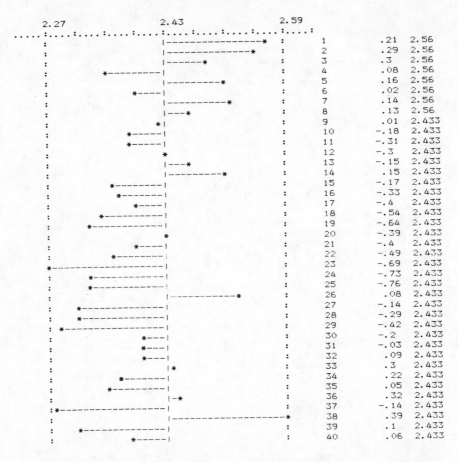

Fig. 3.12 — Results of processing the data in LEADQC.DAT with the REPORT program.

Table 3.7 — Cusum V-Mask estimates of the moving process means in the LEADQC.DAT data when scanned with a V-Mask with '*a*' = 0.1674 and '*b*' = 2.

DATA ITEM Nos.	ARITHMETIC MEAN	MEDIAN	MEDIAN PROCESS MEAN	RATIOS OF THE THREE RESULTS
1–20	9.86	10.15	9.83	1:1.029:0.997
21–27	8.72	8.50	8.28	1:0.975:0.950
28–40	9.11	9.10	8.40	1:0.999:0.922

correct the cause of the negative bias), the process mean remained rock steady and 'perilously' close to the lower action limit of 8.2. Hence a short run of significantly out-of-limits Trigg tracking signals, which is also associated with a sustained shift to a new steady process mean, gives a good indication that a significant trend has become stablized.

In summary, the REPORT program provides four informative tests for between batch quality control:

The standard normal deviate plot — a convenient alternative to the conventional QC chart.

The Trigg tracking signal — an effective device for detecting trends within the QC data.

The cusum V-mask process mean — a device that persists in signalling the resting position of the last trend.

The moving estimate of the standard deviation – a device that provides a reliable local estimate of the SD of the QC data, and thereby reflects the current precision of the assay concerned.

4

Laboratory computing: a discussion

4.1 SOME EXPECTATIONS AND REALITIES

To remain competitive the analytical laboratory must be 'computerized' but it is by no means clear what penetration should be advised or tolerated. As far as the financial activities of a commercial laboratory are concerned, there is little distinction between it and those of any of a large number of service enterprises. However, if a computerized billing and accounting system has been installed it is logical to make some attempts to provide links between it and the analytical-report generation activity.

The purchase of new instrumentation which boasts 'computer compatible' input and output ports on the rear panel then fuels the argument for direct instrument-to-computer link ups. And so the process continues, with the result that there is an 'in house' evolution of computerization which, in some instances, is an outstanding success but in others becomes the scapegoat for a miscellany of mistakes, delays and erratic performance.

An excellent series of articles on laboratory computerization has appeared in *Analytical Chemistry* [34–47] and these are well worth reading. Several of the articles include case studies of organizations that have computerized their laboratories, and these clearly illustrate the evolutionary nature of the activities in this area, with each laboratory having arrived at different and fairly unique solutions.

Computerization would probably have a better public-relations image if it had not become so immersed in its own jargon. However, there are now a number of dictionaries to help the newcomer overcome the language barrier. Another barrier is the poor keyboard skill of the average laboratory scientist or technologist, but this should only be a temporary handicap because there are several typing tutor programs available, particularly for personal computers. The computer keyboard itself has been the subject of justifiable criticism and 'repetition strain injury' is probably the most notorious indictment. It has been pointed out that the inventor of the standard QWERTY keyboard, Christopher L Sholes (1873), purposely designed it to slow down the typist and thereby reduce the risk of jamming the key mechanisms. No such problem exists on an electronic keyboard and an alternative layout would appear to be long overdue.

New alternatives to the keyboard include the 'mouse', the 'touch screen', the 'touch pad' as used in many computer-aided design installations, and a variety of x, y digitizers, voice recognition facilities, and type-reading equipment.

4.2   SOME JUSTIFICATIONS FOR COMPUTERIZATION

A particular task might be computerized for two or more of the following reasons:

(a) It makes a repetitive or tedious task easier to do and less prone to human error.

(b) It is faster than conventional manual or semi-automated methods.

(c) It produces a product that is at least as good if not better than alternative methods.

(d) It promotes better communication and understanding between the producer of the the 'product' and the customer.

(e) It permits more complex data manipulations to be performed routinely.

(f) It makes it unnecessary to enter the same data twice.

The last two reasons deserve some amplification. Advanced analytical techniques frequently require the collection of data at high sampling rates and across a wide dynamic range. Without the aid of a computer these techniques would have remained in the research laboratories, but now, with the addition of a dedicated compact computer, instruments such as quadrupole mass spectrometers, inductively coupled plasmas, X-ray diffraction instruments, NMR spectrometers etc. have been brought into the routine laboratory. Sophisticated algorithms within their internal computer manipulate their complex detector signals into useful information. It is therefore anomalous to witness the operator of such a piece of equipment tearing off the results printout and walking across to another terminal to manually rekey them into the central computerized reporting system. 'Incompatibility' is the usual blanket excuse given; this may be partially valid, but the difficulties could probably be overcome.

We have experimented with the use of a printer spooler to overcome one such 'incompatibility'. The output from a beta counter was available at 300 baud from an RS232 plug on the rear panel. A printer spooler, with 32 K of internal memory was connected to this output. The spooler itself had been modified so that it could be powered from an internal rechargeable battery. When a batch of tubes have been counted the spooler can be unplugged from the mains supply, carried to a computer two rooms away, plugged into the RS232 input of the computer and the contents of the spooler read into the combined curve-fitting and unknowns-calculation program. The additional advantage of the printer-spooler approach has been that the instrument does not have to be polled continuously by the central computer to see if it should break off from its current task and attend to filing the data being transmitted to it down an RS232 line permanently attached to one of its input ports. On-line data acquisition is probably difficult unless the laboratory computer possesses multitasking software. The software required to provide adequate surveillance of just three dissimilar instruments is not simple. Batch processing of analytical data delivered on a computer-readable medium, (printer spooler, floppy disk, magnetic tape, even punched paper tape) is still the easiest method for the average sized laboratory.

Another area where computerization can save effort is in the entry of customer and specimen details. In our clinical laboratory we have introduced 'machine readable' labels which facilitate this process. Bar codes have been implemented on some instruments for automatic specimen identification by the machine; typically a small scanner is located alongside the carousel containing the specimens labelled with bar codes. However, the information on the label is limited to little more than the specimen's accession number because of space limitations. In our laboratory we require repeated access to an extensive amount of textual information: name, analyte, date, customer's name and telephone number, charge code, etc. Some sort of text scanner might provide a partial solution, but would probably be expensive and unable to read handwritten requests. A magnetic strip, as used on bank credit cards, could hold sufficient information to offer a possible solution. This is a problem that deserves urgent attention in many laboratories.

4.3 THE SELECTION OF SUITABLE SOFTWARE

The programs we have provided in this book are straightforward adaptations of accepted quality-control techniques. They should not take long to key in and check for transcription errors. However, the publisher can provide further details on the availability of the programs on floppy disks for a range of microcomputers.

The modular design of these programs allows improvements, expansion or partial transfer across to the reader's existing computerized scheme. However, because they are simple and straightforward they are vulnerable to intentional or unintentional alteration/tampering. It is therefore recommended that once the familiarization period is completed, the programs should be compiled by using a BASIC compiler program. This will render the code immune to tampering and hence consistent program performance can be guaranteed. The data files, however, would still be open to manipulation, so their storage within a protected area of disk space is probably justified. Although we have made no provision in the QCFILER program for analysts to place their initials alongside the QC data that they have entered personally, it is probably a worthwhile modification particularly if a number of staff perform the same assay in the same laboratory.

Proprietary QC software is available and there has been a recent review of several packages [48–49]. For various reasons proprietary software can be cumbersome to use since it may include extensive error checking, and its design aims to cater for all possible contingencies. These features lead to lengthy programs that occupy and/or request access to significant memory space, e.g. two megabytes for Lab Trac QC-2. They can appear to take an unwarranted period of time to execute some seemingly simple tasks. In addition the code, and hence the equations and algorithms used, often are not readily accessible as a listing because of the manufacturer's desire to protect the copyright. This is at the user's expense and is probably not in the best interests of the majority of responsible analysts who have no wish to be involved in piracy.

4.4 MULTIUSER SYSTEMS AND REMOTE OPERATIONS

It appears likely that for true multiuser operation the UNIX operating system will become the accepted standard. In our laboratory we are in the process of implementing a UNIX system running the database program UNIFY, to cater for the demands for on-line data acquisition from our instrumentation and the need for a versatile multiuser database. Activities such as routine wordprocessing and database and statistical analysis of a 'one-off' nature will continue to be catered for on a five-console concurrent CPM–86 system. The consoles will probably be dual-purpose units, capable of being switched between the two minicomputers as required.

In the past, permanent on-line data acquisition in our laboratory was restricted to a 16-well gamma counter linked to a dedicated Hewlett Packard HP 9825. Once the UNIX system has been established this will be extended to include a number of other computer-compatible instruments. The potential for remote data collection from and communication with, analytical instruments is considerable [50]. An instrument — modem — public telephone line — modem — computer system can be used to interrogate instruments in remote laboratories and obtain from them QC and specimen data, and to send back alterations to analytical protocols when necessary.

4.5 CONCLUDING REMARKS

There is considerable scope for improvements to many of the analytical laboratory's activities by means of appropriate computerization. However, before any of these are implemented by an organization, a critical appraisal of the cost and effort *vs.* savings and improvements must be made if staff confidence is to be maintained. The manuals which come with the hardware and software have to be read in spite of their often sterile style and significant length. Considerable time must be allowed for just reading and completing the tutorials that are supplied with the equipment. The general level of computer awareness of all the staff using computerized equipment has to be raised. Each organization should probably nominate two or more staff who will be responsible for gaining some expertise with the equipment, so that they can act both as the 'in house' problem referral point and as the interface with the technical experts. Finally, with the rapid rate of developments in the electronic and computing-components industry, laboratory managers and their accountants must come to terms with high 'hardware redundancy rates'. At present these are quite out of step with their previous experiences and new formulae for equipment capital value depreciation rates have to be derived when computers and their peripherals are being considered for purchase or replacement.

Appendix A
Gaussian data generator

The program included in this Appendix was written to produce a table of data points expressed in standard deviation, SD, units which would follow a Gaussian (i.e. Normal) distribution and which would also appear within the table in an entirely random sequence. "Table Ia : Ordinates of the Normal Probability Curve" in the appendix to Adler and Roessler's book provided the proportional frequencies in each SD interval from the mean, SD = 0, to the extreme, SD = 4.00. We have produced a table of values set at 0.1 SD unit intervals apart, which required us to select the following frequencies of SD's:

SD	=		0	+0.1	+0.2	+0.3	+0.4	+0.5	+0.6	+0.7
Freq	=		40	40	39	38	37	35	33	31
SD	=	+0.8	+0.9	+1.0	+1.1	+1.2	+1.3	+1.4	+1.5
Freq	=	29	27	24	22	19	17	15	13
SD	=	+1.6	+1.7	+1.8	+1.9	+2.0	+2.1	+2.2	+2.3
Freq	=	11	9	8	7	5	4	4	3
SD	=	+2.4	+2.5	+2.6	+2.7	+2.8	+2.9
Freq	=	2	2	1	1	1	1

(Freq = Frequency = Proportional Frequency×100)

In addition, ten outliers were introduced from 3.0 SD to 3.9 SD. These frequencies were included in the DATA statement, line 9064. The program begins by dimensioning an empty array of 528 cells and then filling each of these with a marker value of 10. The co-ordinate of a cell is selected at random, line 9034, and if it contains a '10', the SD currently being dispersed into the array is assigned to that cell; (line

9036). If, however, the cell contains a value other than '10' then it is recognised as a 'filled' cell, and the program continues a random search until it finds an empty cell.

The purpose of Table A1 is to provide a source of simulated experimental data that has an overall random sequence and an underlying Gaussian distribution with a reasonable frequency of outliers, in this instance 10 per 528 data points (= 1.9%). Hence, given a mean value for a measurement, MX, and its SD, there is a simple formula whereby simulated data can be derived:

$$\text{Simulated data point} = MX \pm (SD \times \text{Value from the Table})$$

The decision as to whether or not the right hand term should be added or subtracted can be determined for example by making successive selections from a table of random numbers, e.g. Table A2, and if the value is odd then selecting subtraction and if even, addition, or *vice versa*.

Table A2, which contains 495 random numbers between 0 and 9999, was generated by using the BASIC random number function and the simple program given as a footnote to the Table.

Listing of GAUSSGEN

```
9000 'GAUSSGEN
9005 'GAUSSIAN DATA GENERATOR WITH 10 OUTLIERS PER 528 DATA POINTS
9010 'T F HARTLEY, JULY 1985
9015 DIM GD(528)
9020 INPUT "GIVE ANY NUMBER TO SEED RND NO GENERATOR ";S
9025 RANDOMIZE S
9030 FOR I = 1 TO 528
9035 GD(I) = 10
9040 NEXT I
9045 N = 0
9050 FOR SD = 0 TO 3.9 STEP 0.1
9055 READ F
9060 FOR J = 1 TO F
9065 P = INT(528*RND + 1)
9070 IF GD(P) = 10 THEN GD(P) = SD ELSE 9065
9075 N = N + 1 : PRINT SD; N;
9080 NEXT J
9085 NEXT SD
9090 PRINT "ARRAY FILLED"
9095 T = 0
9100 LPRINT : LPRINT : LPRINT
9105 LPRINT "GAUSSIAN DATA : 528 DATA POINTS IN +VE SD UNITS WITH 10
     "; : LPRINT " OUTLIERS >= 3 SD" : LPRINT
9110 FOR I = 1 TO 528
9115 LPRINT USING "#.#"; GD(I); :LPRINT " ";
9120 T = T + 1
9125 IF T = 14 THEN T = 0 : LPRINT
```

```
9130 NEXT I
9135 |PRINT
9140 DATA 40, 40, 39, 38, 37, 35, 33, 31, 29, 27, 24, 22, 19, 17,
15, 13, 11, 9, 8, 7, 5, 4, 4, 3, 2, 2, 1, 1, 1, 1, 1, 1, 1,
1, 1, 1, 1, 1, 1, 1
```

Table A1 — Random data points corresponding to the positive half of a Gaussian distribution

0.3	0.5	0.6	1.1	1.0	1.1	2.0	0.7	0.6	0.9	1.4	1.2	0.7	0.2
0.2	0.8	1.9	1.2	0.0	1.2	0.3	1.7	2.3	0.9	1.9	2.8	0.7	1.0
2.0	0.1	2.2	0.5	0.2	0.0	1.5	0.5	1.3	3.3	0.6	1.0	1.4	0.9
0.6	0.3	1.0	1.2	1.3	1.3	0.2	3.5	0.2	0.4	1.7	0.3	0.0	0.1
0.1	2.2	0.8	0.5	0.9	0.1	1.5	0.4	0.8	0.2	1.3	1.0	1.4	0.2
0.3	0.3	3.6	0.0	0.1	1.2	0.2	0.0	0.8	1.0	1.6	1.0	0.4	1.6
0.8	1.4	0.3	0.1	2.5	0.7	1.0	0.3	0.4	0.1	0.8	0.7	0.0	0.5
0.8	0.5	1.1	3.1	0.9	0.7	0.3	3.5	0.3	1.3	1.4	0.1	0.1	0.0
0.2	0.2	0.5	0.9	0.2	1.6	0.7	0.3	3.9	2.6	0.4	0.8	1.9	0.5
0.7	3.4	1.3	0.6	0.1	0.3	1.1	0.0	0.1	2.7	1.1	0.6	0.6	3.0
1.5	0.2	1.3	1.1	1.5	0.4	0.2	0.8	0.1	0.5	1.7	0.2	1.0	0.4
0.8	1.5	0.2	0.2	0.7	0.3	0.2	0.3	0.2	0.0	1.1	0.0	1.2	1.4
1.2	1.6	1.7	0.6	0.4	0.0	0.4	0.1	0.4	0.8	0.8	0.9	1.1	0.7
0.2	0.8	0.6	0.8	0.4	1.0	2.1	1.4	1.7	1.8	1.6	1.3	0.8	1.0
1.2	0.4	1.0	0.1	0.1	0.8	0.6	3.7	0.0	1.0	1.2	0.4	1.2	0.1
0.2	0.2	0.0	1.0	0.4	0.1	0.8	0.1	1.7	1.6	0.2	0.0	2.1	1.0
0.0	0.9	1.7	0.7	0.6	0.9	1.4	0.4	0.1	1.9	0.5	1.0	0.3	0.4
0.5	1.6	1.8	1.6	0.5	0.9	0.6	0.7	0.8	0.6	2.0	0.3	0.1	1.2
0.3	0.7	1.3	0.4	0.0	0.1	0.2	1.2	0.8	0.6	0.8	1.9	0.3	2.3
0.3	1.0	0.2	1.4	2.4	0.9	0.3	1.2	0.4	0.0	0.8	1.4	0.7	0.9
0.7	1.2	0.4	0.5	0.0	0.9	0.2	0.4	0.2	0.6	0.5	0.1	1.3	0.9
0.4	0.4	1.1	0.1	0.9	1.5	2.0	0.3	0.1	0.6	1.1	0.1	0.5	0.2
1.5	0.1	1.8	0.5	1.1	0.7	0.8	1.0	0.0	2.2	0.0	0.1	0.6	0.4
0.9	0.4	0.6	0.1	0.3	0.0	0.4	0.7	1.5	1.1	0.7	1.8	0.5	0.7
0.0	0.1	3.8	0.5	1.8	1.0	0.2	0.0	1.1	0.0	2.3	0.4	0.3	0.3
0.6	1.4	1.3	1.3	0.4	0.0	1.5	0.3	0.4	2.0	0.5	0.6	0.2	1.6
0.8	0.3	0.5	0.0	0.2	0.1	0.1	0.7	1.4	0.7	1.6	0.7	1.2	0.4
2.1	0.3	0.2	1.8	0.5	0.3	0.4	0.0	1.2	0.6	1.4	0.0	1.0	0.9
0.5	0.9	0.9	0.5	0.1	0.0	1.1	0.6	1.3	0.5	0.1	1.8	0.7	0.9
0.3	0.8	0.0	0.7	0.9	1.9	0.1	0.8	1.7	0.2	0.5	0.9	0.7	0.7
0.3	1.4	0.4	0.3	0.6	0.4	1.2	1.9	0.5	0.3	1.1	1.1	0.1	1.1
1.1	0.9	0.2	0.6	0.6	0.6	0.4	1.1	1.3	1.3	1.2	0.7	0.6	0.8
1.3	0.2	2.9	1.1	1.7	0.5	1.3	0.0	0.8	0.5	2.1	0.1	0.0	0.6
0.3	2.2	0.6	0.7	1.5	1.5	1.5	0.8	0.0	0.1	0.4	0.2	0.0	2.4
0.3	0.2	0.4	1.0	3.2	1.8	0.3	0.2	0.7	1.1	0.9	0.6	0.5	0.5
1.1	1.4	0.5	0.0	0.1	0.6	0.5	1.0	0.9	0.7	0.3	0.4	0.7	0.0
1.5	0.4	0.0	0.0	0.6	0.2	0.5	1.0	0.2	0.1	0.9	0.9	0.0	2.5
0.0	1.0	0.3	0.6	1.2	0.1	0.5	0.7	0.3	1.6				

Table A2 — Random numbers

4971	8208	1049	5795	3318	8555	9902	7652	6693	8948	6348
8420	8503	833	9628	1336	8474	7172	538	4201	2495	408
5020	6094	9958	7935	5256	1592	8609	5968	8261	2489	6883
5509	1980	7013	3755	9319	8878	6549	6080	8384	7034	9854
3437	9387	8740	4776	473	4272	6586	7474	6749	8216	7061
9136	4081	5100	1644	198	1384	471	5399	2065	7016	2016
3065	6324	8624	7531	2655	1470	484	387	6547	7117	8348
2499	75	5723	1786	4796	7630	1808	4029	5129	6247	8669
9666	5111	6630	5823	3350	2093	3393	4572	6144	426	3606
8323	5115	4600	7184	3360	3142	1521	4773	7268	5631	7846
3221	117	4738	4835	8545	3066	1084	692	6401	4447	2977
405	52	3368	3129	3095	5839	798	2874	6856	6146	8169
3890	1104	9872	3463	7413	6412	397	4809	960	1863	1052
1577	2546	9888	8091	2969	2401	8610	1153	4119	2193	1692
1236	9677	1743	5675	6165	4913	3244	4827	3558	4302	2984
3484	1298	7625	559	5892	5500	3559	5593	3554	5981	768
1962	74	9226	6504	9507	3498	8861	2694	3252	6160	3618
9539	129	2023	2806	6603	3192	1527	9277	1409	3336	3928
9899	5536	8516	7146	3521	8204	468	5106	6444	7378	8926
2583	5389	8356	3339	7545	7964	9861	2092	997	637	7640
9227	4325	4200	5556	8655	4141	6742	4178	5882	4623	4112
6881	7223	7362	7897	3433	6522	8706	879	1096	9373	4600
15	3213	4834	2936	6898	5306	4651	3390	5729	8669	9972
4770	8786	1460	5571	9859	9389	7296	1561	7757	2376	1349
2617	3729	2065	3934	4693	6542	7724	1608	9241	3315	6544
5764	4930	1860	4102	8718	6865	527	2333	7313	3303	5544
5000	4002	9877	5867	658	5451	1310	5518	3128	8626	636
3432	9463	4392	5918	801	5416	8602	4469	5913	8412	694
4566	6504	8011	9784	4672	9594	3885	8258	157	8835	1279
2870	7510	9353	3985	3985	3330	6065	93	4805	8405	8402
7216	3661									
3529	9497	1135	3671	2393	1766	993	9584	9056	2693	4677
9056	7065	8191	4010	7781	3824	6194	1471	6831	6063	7751
4409	7505	2807	5620	6766	3669	5845	362	8296	1602	7700
3935	1281	7339	786	4750	1514	1916	6447	6201	1704	
9701	3627	817	2273	5586	5806	8183	1719	7787	8332	6510
7509	562	7009	4349	8434	4686	4195	9813	4443	5413	2394
7294	3993	7951	5351	1679	8890	426	6644	77	6634	4355
5779	1029	260	2610	1784	1230	5686	4889	4302	9751	6555
688	2185	9752	6576	9764	9588	7705	9372	106	9562	110
5821	3407	5148	7717	8901	8027	7971	7242	6034	9466	4335
584	8927	8332	8814	1551	921	5808	5348	70	1828	6622
5836	3912	281	9517	6229	7909	3575	5217	125	669	5009
1897	9864	8802	8734	8154	5116	8632	2092	3471	8982	5420
5463	6190	1583	82	9610	8980	1156	3404	517	1802	2647

Footnote:
```
 5 T=0
10 For I=1 TO 500
20 X=RND(0)
25 PRINT X; " ";
26 X=INT(10000*X+0.5)
30 LPRINT TAB(T) X; : T=T+6
35 IF T=65 THEN LPRINT : T=0
40 NEXT I
```

Appendix B
BASIC version of the SPLINE program [1]

```
10 'SPLINE PROGRAM FROM [1]
20 'TRANSLATED INTO BASIC BY T. HARTLEY
30 'APPENDIX B
40 'COMPUTERIZED QUALITY CONTROL
50 'PUBLISHED BY ELLIS HORWOOD, ENGLAND, 1986
60 '
70 WIDTH 80
80 DIM XI(18), C(4, 18), D(18), DIAG(18)
 90 N1 = 7
100 N2 = 11
110 '
120 FORJ= 1 TO 7
130 READ XI (J), C(1,J)
140 NEXT J
150 '
160 READ C(2,1) : ' = 0.8
170 READ C(2,7) : ' = 0.2
180 DIAG(1) = 1 : D(1) = 0 : S = 2 : N = 7
190 GOSUB 510 : 'SPLINE
200 GOSUB 750 : 'CALCCF
210 PRINT "DONE 1ST MATRIX"
220 '
230 FOR J=8 TO 18
240 READ XI (J), C(1,J)
250 NEXT J
260 '
270 READ C(2,8) : ' -1.80
280 READ C(2, 18) : ' = -0.35
290 '
300 DIAG(8) = 1 : D(8) = 0 : S = 9 : N = N1 + N2
310 '
320 GOSUB 510 : 'SPLINE
330 GOSUB 750 : ' CALCCF
340 PRINT "DONE 2ND MATRIX"
```

```
350 '
360 FOR K=1 TO 5 : LPRINT : NEXT K
370 FOR X = 0 TO 18 STEP 0.5
380 GOSUB 870 : 'PCUBIC
390 FX = PCUBIC
400 PRINT I, X, FX;
410 T = INT( 10 * (FX − 2.5 + 0.5) )
420 PRINT TAB(35 + T ) "*"
430 LPRINT TAB (T) "*"; TAB(50) X; TAB(60) FX
440 NEXT X
450 '
460 PRINT "DONE************************" : STOP
470 '
480 DATA 0, 2.51, 1, 3.30, 2, 4.04, 3, 4.70, 4, 5.22, 5,
    5.54, 6.1, 5.80, 0.8, 0.2
490 DATA 6.1, 5.80, 6.3, 5.55, 6.5, 5.44, 6.7, 5.40, 7, 5.40
    −1.80, −0.35
500 '
510 'SPLINE SUBROUTINE
520 '
530 FOR M=S TO N
540 D(M) = XI(M) − XI(M−1)
550 DIAG(M) = (C(1,M) − C(1,M−1)) / D(M)
560 NEXT M
570 '
580 FOR M=S TO N−1
590 C(2,M) = 3 * ( D(M) * DIAG(M+1) + D(M+1) * DIAG(M) )
600 DIAG(M) = 2 * ( D(M) + D(M+1) )
610 NEXT M
620 '
630 FOR M = S TO N−1
640 PRINT "* "; : G = − D(M+1) / DIAG(M−1)
650 DIAG(M) = DIAG(M) + G * D(M−1)
660 C(2,M) = C(2,M( + G * C(2,M−1)
670 NEXT M
680 '
690 FOR M=N−1 TO S STEP −1
700 C(2,M) = ( C(2,M( − D(M) * C(2,M+1)) / DIAG(M)
710 NEXT M
720 '
730 RETURN
740 '
750 'CALCCF SUBROUTINE
760 '
770 FOR I = S − 1 TO N − 1
780 DX = XI(I+1) − XI(I)
790 DIVDF1 = ( C(1, I+1) − C(1,I) ) / DX
800 DIVDF3 = C(2,I) + C, I+1) − 2 * DIVDF1
810 C(3,I) = ( DIVF1 − C(2,I) − DIVDF3) / DX
820 C(4,I) = DIVDF3 / DX ↑ 2
830 NEXT I
840 '
850 RETURN
860 '
870 PCUBIC
880 FOR J = 1 TO 17
890 IF X >= XI(J) AND X < XI(J+1) THEN GOTO 920
900 NEXT J
910'
920 DX = X − XI(J)
930 I = J
940 PCUBIC = C(1, I) + DX * ( C(2, I) + DX * (C(3,I) + DX*C(4,I)))
950 RETURN
```

Appendix C
Program for generating the operating characteristic table for batch quality control

```
9200 'OCTABLE.BAS
9202 'OPERATING CHARACTERISTIC CURVE
9204 'T F HARTLEY
9206 'COMPUTERIZED QUALITY CONTROL
9208 'PUBLISHED BY ELLIS HORWOOD, ENGLAND, 1986
9210 '
9212 '
9214 DIM PA(9,12)
9216 WIDTH LPRINT 100
9218 '
9220 FOR I = 1 TO 10 : PRINT : NEXT I
9222 PRINT STRING$(60, "*")
9224 PRINT "OPERATING CHARACTERISTIC TABLE FOR BATCH QC DESIGN"
9226 PRINT "ACCEPTANCE CRITERION IS THAT ALL THE QC's IN THE BATCH"
9228 PRINT "MUST FALL WITHIN THE PRESCRIBED LIMITS = ± 2 SD'9230 PRINT "NO
OUTLIERS ARE PERMITTED"
9232 PRINT STRING$(60,"*")
9234 PRINT
9236 X = 1
9238 '
9240 FOR C = 1 TO S
9242 READ P
9244 NEXT C
9246 '
9248 FOR J = S TO 8
9250 READ X
9252 NEXT J
9258 '
9260 FOR R = 1 TO 12
9262 READ N
9264 PA = ( 1 − P ) ↑ N
9266 PA(S,R( = ( INT( PA * 1000 + 0.5)) / 10
9268 NEXT R
9270 '
```

```
9272 S = S + 1
9274 RESTORE
9276 '
9278 IF S > 9 THEN GOTO 9288 ELSE 9240
9280 '
9282 DATA 0.001, 0.01, 0.02, 0.05, 0.1, 0.2, 0.3, 0.4, 0.5
9284 1, 2, 3, 4, 5, 10, 15, 20, 30, 40, 50, 100
9286 '
9288 LPRINT : LPRINT
9290 LPRINT "OPERATING CHARACTERISTIC TABLE FOR BATCH QC DESIGN"
9292 LPRINT STRINGS$ (80, "–")
9294 LPRINT "ACCEPTANCE CRITERION IS THAT ALL QC's IN THE BATCH"
9296 LPRINT "MUST FALL WITHIN THE PRESCRIBED LIMITS = ± 2 SD"
9298 LPRINT "NO OUTLIERS ARE PERMITTED"
9300 LPRINT "FIGURES IN THIS TABLE EQUAL THE PERCEWNTAGE CHANCE"
9302 LPRINT "OF ACCEPTING THE BATCH"
9304 LPRINT : LPRINT : LPRINT "                         # OF QC's";
9306 LPRINT TAB(20) "PROPORTION DEFECTIVE IN THE BATCH, %"
9308 LPRINT : LPRINT " ";
9310 RESTORE
9312 T = 15
9314 '
9316 FOR C = 1 TO 9
9318 READ X
9320 LPRINT TAB(T( 100 * X ;
9322 T = T + 7
9324 NEXT C
9326 '
9328 LPRINT : LPRINT
933 '
9332 FOR R = 1 TO 12
9334 '
9336 READ X
9338 LPRINT "          "; X;
9340 '
9342 T = 15
9344 FOR C = 1 TO 9
9346 LPRINT TAB(T) PA(C,R);
9348 T = T +7
9350 NEXT C
9352 '
9345 LPRINT
9356 NEXT R
9358 '
9360 STOP
```

References

[1] S. D. Conte and C. De Boor, *Elementary Numerical Analysis. An Algorithmic Approach,* 3rd Ed., McGraw-Hill, New York, 1980.

[2] C. H. Reinsch, *Numerische Mathematik,* 1967, **10,** 177.

[3] G. W. Snedecor and W. G. Cochran, *Statistical Methods,* 7th Ed., Iowa State University Press, Iowa, 1980.

[4] J. E. A. McIntosh and R. P. McIntosh, *Mathematical Modelling and Computers in Endocrinology,* Springer-Verlag, Berlin, 1980.

[5] F. J. Hawker and G. S. Challand, *Clin. Chem.,* 1981, **27,** 14.

[6] D. Rodbard, *J. Clin. Endocrinol. Metab.,* 1968, **29,** 352.

[7] M. J. R. Healy, *Biochem. J.* 1972, **130,** 207.

[8] D. J. Finney, *Biometrics,* 1976, **32,** 721.

[9] *Radioimmunoassay and Related Procedures in Medicine,* Vol. 1, IAEA, Vienna, 1978, p. 339.

[10] C. G. Paradine and B. H. P. Rivett, *Statistical Methods for Technologists.* 3rd Ed., The English Universities Press, London, 1960, p. 140.

[11] M. J. Maroney, *Facts from Figures,* 3rd Ed. Penguin, Harmondsworth, 1968. p. 173.

[12] P. M. G. Broughton and L. Eldjarn, *Ann. Clin. Biochem.,* 1985, **22,** 625.

[13] L. Eldjarn and P. M. G. Broughton, *Ann. Clin. Biochem.,* 1985, **22,** 635.

[14] A. Wald, *Ann. Math. Stats.,* 1940, **11,** 284.

[15] W. E. Deming, *Statistical Adjustment of Data,* Wiley, New York, 1943, p. 184.

[16] M. S. Bartlett, *Biometrics,* 1949, **5–6,** 207.

[17] E. S. Keeping, *Biometrics,* 1956, **12,** 445.

[18] J. Mandel and F. J. Linnig, *Anal. Chem.*, 1957, **29**, 743.

[19] R. N. Barnett and W. J. A. Youden, *Am. J. Clin. Pathol.*, 1970, **54**, 454.

[20] R. R. Sokal and F. J. Rohlf, *Biometry. The Principles and Practice of Statistics in Biological Research.* 2nd Ed., Freeman, New York, 1981.

[21] T. F. Hartley, J. C. Philcox, J. F. Beesley and J. B. Robinson, *J. Food Nutr.*, 1986, **42** (4), in the press.

[22] C. D. Lewis, *Qual Assurance*, 1980, **6**, 3.

[23] P. J. Harrison and O. L. Davies, *Oper. Res.*, 1964, **12**, 325.

[24] R. A. Woodward and P. L. Goldsmith, *Cumulative Sum Techniques*, Oliver and Boyd, Edinburgh, 1964.

[25] D. W. Trigg, *Oper. Res. Q.*, 1964, **15**, 271.

[26] C. E. Hope, C. D. Lewis, I. R. Perry and A. Gamble, *Brit. J. Anaesth.*, 1973, **45**, 440.

[27] O. L. Davies and P. L. Goldsmith, *Statistical Methods in Research and Production*, 4th Ed., Longman, London, 1972.

[28] W. D. Ewan and K. W. Kemp, *Biometrika*, 1960, **47**, 239.

[29] K. W. Kemp, *J. R. Stat. Soc. B*, 1961, **23**, 149.

[30] R. W. H. Edwards, *Ann. Clin. Biochem.* 1980, **17**, 205.

[31] R. G. Brown, *Smoothing, Forecasting and Prediction of Discrete Time Series*, Prentice Hall, Englewood Cliffs, 1962.

[32] G. S. Cembrowski, J. O. Westgard, A. A. Eggert and E. C. Toren, *Clin. Chem.*, 1975, **21**, 1396.

[33] M. Batty, *Oper. Res. Q.*, 1969, **20**, 319.

[34] R. E. Dessy, *Anal. Chem.*, 1982, **54**, 1167A.

[35] R. E. Dessy, *Anal. Chem.*, 1982, **54**, 1295A.

[36] R. E. Dessy, *Anal. Chem.*, 1983, **55**, 70A.

[37] R. E. Dessy, *Anal. Chem.*, 1983, **55**, 277A.

[38] R. E. Dessy, *Anal. Chem.*, 1983, **55**, 650A.

[39] R. E. Dessy, *Anal. Chem.*, 1983, **55**, 756A.

[40] R. E. Dessy, *Anal. Chem.*, 1984, **56**, 725A.

[41] R. E. Dessy, *Anal. Chem.*, 1984, **56**, 855A.

[42] R. E. Dessy, *Anal. Chem.*, 1985, **57**, 77A.

[43] R. E. Dessy, *Anal. Chem.*, 1985, **57**, 310A.

[44] R. E. Dessy, *Anal. Chem.*, 1985, **57**, 693A.

[45] R. E. Dessy, *Anal. Chem.*, 1985, **57**, 805A.

[46] R. E. Dessy, *Anal. Chem.*, 1986, **58**, 78A.

[47] R. E. Dessy, *Anal. Chem.*, 1985, **58**, 313A.

[48] C. E. Stewart and B. S. Oxford, *J. Med. Technol.* 1985, **2**, 621.

[49] B. S. Oxford, *J. Med. Technol.* 1985, **2**, 629.

[50] P. Scott, D. J. Andrews and P. Gosling, *Isr. J. Clin. Biochem. Lab. Sci.*, 1985, **4**, 108.

Supplementary reading

H. L. Alder and E. B. Roessler, *Introduction to Probability and Statistics,* 6th Ed., Freeman, San Francisco, 1976.

A. S. Blum, Computer Evaluation of Statistical Procedures and a New Quality Control Statistical Procedure, *Clin. Chem.,* 1985, **31,** 206.

R. Caulcutt and R. Boddy, *Statistics for Analytical Chemists,* Chapman and Hall, London, 1983.

G. S. Cembrowski and J. O Westgard, Quality Control of Multichannel Haematology Analyzers: Evaluation of Bull's Algorith. *Am. J. Clin. Pathol.,* 1985, **83,** 337.

G. S. Challand. Automated Calculation of Radioimmunoassay Results, *Ann. Clin. Biochem.* 1978, **15,** 123.

R. P. Channing Rodgers, Data Analysis and Quality Control of Assays: A Practical Primer, in *Practical Immunoassay. The State of the Art,* W. R. Batt, ed., Dekker., New York, 1984.

P. J. Cornbleet and N. Goochman, Incorrect Least Squares Regression Coefficients in Method Comparison Studies, *Clin. Chem.,* 1979, **25,** 432.

M. F. Delaney, Chemometrics, *Anal. Chem.,* 1984, **56,** 261R.

P. England and O. Cain, A Simple Method for the Interpolation of Radioimmunoassay Data, *Clin. Chim. Acta.* 1976, **72,** 241.

E. L. Grant and R. S. Leavenworth, *Statistical Quality Control,* 5th Ed., McGraw-Hill, New York, 1980.

J. R. Green and D. Margerison, *Statistical Treatment of Experimental Data,* Elsevier, Amsterdam, 1978.

T. Groth, H. Falk and J. O. Westgard, An Interactive Computer Simulation Program for the Design of Statistical Control Procedures in Clinical Chemistry, *Comput. Programs Biomed.* 1981, **13,** 73.

J. D. F. Habbema and G. J. Van Oortmarssen, Performance Characteristics of Screening Tests, *Clinics Lab. Med.,* 1982, **2,** 639.

T. F. Hartley and T. W. Huber, Computing Forms of the Equations for Linear Regression Analysis, *Lab. Pract.,* 1984, **33** (10), 119.

T. F. Hartley and T. W. Huber, Computerised Quality Control with Trigg Trend Detection and Cusum Capabilities, *Tr. Anal. Chem.,* 1984, **3,** 251.

S. L. Jeffcoate and R. E. G. Das, Interlaboratory Comparison of Radioimmunoassay Results. Variations Produced by Different Methods of Calculation, *Ann. Clin. Biochem.* 1979, **14,** 258.

B. Kratochvil, D. Wallace and J. K. Taylor, Sampling for Chemical Analysis, *Anal. Chem.* 1984, **56,** 113R.

C. D. Lewis, Statistical Monitoring Techniques, *Med. Biol. Eng.,* 1971, **9,** 315.

D. L. Massart, A. Dijkstra and L. Kaufman, *Evaluation and Optimization of Laboratory Methods and Analytical Procedures,* Elsevier, Amsterdam, 1978.

R. Meddis, *Statistical Handbook for Non-Statisticians.* McGraw-Hill (UK), London, 1975.

J. C. Miller and J. N. Miller, *Statistics for Analytical Chemistry,* Ellis Horwood, Chichester, 1984.

M. Motta and A. Degli Esposti, A Computer Program for Mathematical Treatment of Data in Radioimmunoassay, *Comput. Programs Biomed.* 1981, **13,** 121.

M. Pandin, Computer Program for Quality Control, Especially Suited for the Small Laboratory, (Program for HP-86 Computer). *Clin. Chem.,* 1985, **31,** 166.

L. G. Parratt, *Probability and Experimental Errors in Science.* Wiley, New York, 1961.

A. Pilo and G. C. Zucchelli, Automatic Treatment of Radioimmunoassay Data: An Experimental Validation of Results, *Clin. Chim. Acta,* 1975, **64,** 1.

D. Rodbard and Y. Feldman, Kinetics of Two-Site Immunoradiometric ('Sandwich') Assays. I — Mathematical Models for Simulation, Optimization and Curve Fitting, *Immunochemistry,* 1978, **15,** 71.

D. Rodbard, Y. Feldman, M. L. Jaffe and L. E. M. Miles, Kinetics of Two-Site Immunoradiometric ('Sandwich') Assays. II — Studies on the Nature of the 'High Dose Hook Effect', *Immunochemistry,* 1978, **15,** 77.

E. Samols and E. H. Barrows, Auomated Data Processing and Radioassyas, *Sem. Nucl. Med. VIII*, 1978 (2), 163.

I. Schoen, E. Custer, G. Graham, Z. Bandi and M. H. Surovik, Quality Control Log with Cusum and Clinically Useful Limits Criteria, *Arch. Pathol. Lab. Med.* 1985, **109,** 333.

S. Schwarz, Hewlett Packard 41 Software Package for Radioimmunoassay Data Evaluation and Continuous Batch-to-Batch Quality Control, *Clin. Chem.,* 1985, **31,** 488.

M. R. Spiegel, *Theory and Problems of Statistics. SI (metric) Edition.* McGraw-Hill, New York, 1972.

J. J. Tiede and M. Pagano, The Application of Robust Calibration to Radioimmunnoassay, *Biometrics,* 1979, **35,** 567.

W. Vogt, P. Sandel, Ch. Langfelder and M. Knedel, Performance of Various Mathematical Methods for Computer-Aided Processing of Radioimmunoassay Results, *Clin. Chim. Acta.* 1978, **87,** 101.

J. O. Westgard and T. Groth, Design and Evaluation of Statistical Control Procedures: Applications of a Computer "Quality Control Simulator" Program, *Clin. Chem.,* 1981, **27,** 1536.

J. O. Westgard and M. R. Hunt, Use and Interpretation of Common Statistical Tests in Method Comparison Studies, *Clin. Chem.,* 1973, **19,** 49.

Index